Plasma: From Materials to Emerging Technologies

Plasma: From Materials to Emerging Technologies

Editor

Mirosław Dors

MDPI • Basel • Beijing • Wuhan • Barcelona • Belgrade • Manchester • Tokyo • Cluj • Tianjin

Editor
Mirosław Dors
Polish Academy of Sciences
Poland

Editorial Office
MDPI
St. Alban-Anlage 66
4052 Basel, Switzerland

This is a reprint of articles from the Special Issue published online in the open access journal *Applied Sciences* (ISSN 2076-3417) (available at: https://www.mdpi.com/journal/applsci/special_issues/plasma_materials_emerging_technologies).

For citation purposes, cite each article independently as indicated on the article page online and as indicated below:

LastName, A.A.; LastName, B.B.; LastName, C.C. Article Title. *Journal Name* **Year**, *Volume Number*, Page Range.

ISBN 978-3-0365-2125-1 (Hbk)
ISBN 978-3-0365-2126-8 (PDF)

Cover image courtesy of Bartosz Hrycak.

Contents

About the Editor

Mirosław Dors (Professor) received his M.S. degree in Chemistry from Gdańsk University of Technology, Poland, in 1993, a Ph.D. degree from the Institute of Fluid-Flow Machinery, Polish Academy of Sciences, Gdańsk, Poland, in 2000, and his D.Sc. from Warsaw University of Technology, Poland, in 2009. He is currently an Associate Professor and the Head of the Centre for Plasma and Laser Engineering, Institute of Fluid Flow Machinery, Polish Academy of Sciences, Gdańsk, Poland. He has been working in the areas of plasma physics and chemistry, DC, pulsed and microwave discharges, and plasma applications for environmental technologies. He has co-authored approximately 80 published refereed papers and over 150 conference papers on these topics.

Editorial

Plasma: From Materials to Emerging Technologies

Mirosław Dors

Institute of Fluid Flow Machinery, Polish Academy of Sciences, 80231 Gdańsk, Poland; mdors@imp.gda.pl

Citation: Dors, M. Plasma: From Materials to Emerging Technologies. *Appl. Sci.* **2021**, *11*, 8311. https://doi.org/10.3390/app11188311

Received: 19 August 2021
Accepted: 6 September 2021
Published: 8 September 2021

Publisher's Note: MDPI stays neutral with regard to jurisdictional claims in published maps and institutional affiliations.

1. Introduction

Interest in plasma as a tool in various technological processes has been growing for several decades. This is because of the special advantage of plasma, which is the immediate generation of chemically active radicals. There are also other advantages of plasma, which depend on its source, e.g., low or high temperature (dielectric barrier discharge vs. plasmatrons), large or small volume (electron beam chambers vs. microplasma), high or low homogeneity (low pressure radio-frequency plasma vs. corona discharge), etc. It is no wonder that plasma is used in so many areas, starting with the synthesis of ozone initiated by Werner von Siemens in 1857, through the activation of material surfaces and flow control by actuators and electrohydrodynamic pumps, to the latest applications related to medicine, environmental protection and stopping climate change.

The objective of this Special Issue is to collect reports on the design and characterization of plasma methods which are or can be used in various types of technologies, especially those that solve contemporary problems regarding materials, energy and the environment.

2. Review of Issue Contents

The Special Issue is composed of seven papers covering numerical and experimental research on different aspects related to plasma applications. Here they are introduced in five specific categories that emerge after their reading.

2.1. Plasma Sources

Martines et al. [1] designed a helical resonator for radio-frequency plasma generation starting from a theoretical model. The helical resonator is a concept for the production of high voltage at radio frequency, useful for electrical discharges in gases and plasma sustainment. When properly designed, the helical resonator enables the avoidance of the use of a matching network. In this work, researchers consider the treatment of the helical resonator, including a grounded shield, as a transmission line with a shorted end and an open one, with the latter connected to a capacitive load. The input voltage is applied to a tap point located near the shorted end. The authors derived formulas that allowed a prediction of the resonator performance, consequently enabling one to properly design a plasma source according to the parameters required for the plasma. What is important is that the discussion presented in the paper considers the resonator without plasma: the effects of a plasma formed inside the device is going to be the object of a forthcoming publication.

Luo et al. [2] developed a practical method for controlling the switch been the symmetric and asymmetric modes of a dielectric barrier discharge at atmospheric pressure by changing the frequency of the applied voltage. In this work, through a qualitatively validated 1D fluid model, the discharge evolution, manipulating process and underlying mechanism are also presented. It must be noted that the work is a numerical one. The concept is discussed based on simulations. They showed that the effectiveness of the control strategy was determined by the seed electron level at the frequency-altered phase and that there was a critical range of the seed electron density. For example, under the original driving frequency of 14 kHz, the seed electron level approximately ranges from 2×10^{13} m^{-3} to 8×10^{15} m^{-3}. Researchers found, numerically, that the discharges with an initial driving frequency of 14 kHz could always be converted to the symmetric mode when the control frequency was beyond 30 kHz.

2.2. Plasma Diagnostics

Lee et al. [3] investigated the Coulomb-focused bremsstrahlung spectrum resulting from the electron-atom bremsstrahlung process in nonthermal plasma. They derived the universal expression of the electron-atom bremsstrahlung cross section by using the Thomas–Fermi model with the effective charge method, as well as the effective Coulomb focusing factor for the abovementioned process by using the modified Elwart–Sommerfeld factor with the mean effective charge for the binary-encounter. The obtained results should be useful for diagnostics of radiation processes in nonthermal plasma, for hard-photon X-ray spectroscopy, and for diagnostics of fusion plasmas, magnetized plasmas and dusty plasmas.

2.3. Synthesis of Carbon Structures

Sobczyk and Jaworek [4] studied the process of growth of various carbon structures in low-current high-voltage electrical microdischarges in argon at atmospheric pressure with an admixture of cyclohexane as the carbon source. The following various types of microdischarges generated at this pressure were tested for both polarities of the supply voltage with regard to their applications to different carbon deposit syntheses: Townsend discharge, pre-breakdown streamers, breakdown streamers and glow discharge. The discharge was generated between a stainless-steel needle and a plate made of a nickel alloy with inter-electrode distances varying between 1 and 15 mm. The results of the experiments carried out at different discharge currents, electrode polarities, inter-electrode distances and cyclohexane concentrations brought the authors to the conclusion that the process of carbon deposition (in the form of carbon nanowalls and carbon microfibers), in particular the morphology of deposits and their growth rate, could be successfully controlled but that the optimal conditions could only be determined experimentally. The paper contains numerous scanning electron microscopy (SEM) pictures of deposits and spectra resulting from the optical emission spectroscopy of plasma formed at different discharge currents.

Li et al. [5] applied microwave plasma at atmospheric pressure to synthesize multiwalled carbon nanotubes on stainless steel by chemical vapor deposition (CVD) using ethanol as a precursor. They compared this method with other low-pressure CVD techniques such as conventional CVD, direct current plasma CVD, radio-frequency plasma CVD and microwave plasma CVD. The obtained results showed that carbon tubes synthesized using atmospheric pressure microwave plasma were characterized by a higher growth rate and lower defects concentration when compared to the conventional CVD, but with similar properties to those produced using other plasma methods. Thus, the advantage of the method applied by Li et al. is that it uses atmospheric pressure instead of reduced pressure. The paper contains pictures of produced carbon structures obtained from SEM and a high-resolution transmission electron microscope as well as Raman and X-ray photoelectron spectroscopy (XPS) spectra.

2.4. Plasma for Surface Processing

Krumpolec et al. [6] presented a method for the surface processing of complex polymer-metal composite substrates. Atmospheric-pressure plasma etching in pure hydrogen, nitrogen, their mixture as well as in air was used to fabricate flexible transparent composite poly (methyl methacrylate) (PMMA)-based polymer film with Ag-coated Cu metal wire mesh substrates (with conductive connection sites) by the selective removal of the thin (~10–100 nm) surface PMMA layer. To simulate large-area roll-to-roll processing, the authors used an advanced alumina-based concavely curved electrode generating a thin and high-power density cold plasma layer by the diffuse coplanar surface barrier discharge. After a series of experiments with various gases, it turned out that a short 1 s exposure to pure hydrogen plasma led to a highly selective etching of the surface PMMA film without any destruction of the Ag-coated Cu metal wires embedded in the PMMA structure. The researchers concluded that the applied direct current surface barrier discharge could be used for the fast, large-area, roll-to-roll and selective plasma etching of complex materials like flexible photovoltaic substrates.

2.5. Plasma for Exhaust Gas Cleaning

Cai et al. [7] experimentally studied the process of nitrogen monoxide oxidation (NO) by an atmospheric pressure dielectric barrier discharge (DBD). The subject of the investigation was a gas mixture simulating diesel exhaust from a marine engine. Researchers focused on the effect of electrode parameters such as length, diameter, material and shape. An interesting result was obtained when varying the electrode diameter. Increasing this parameter resulted in a higher oxidation degree of NO and reduced energy consumption. An increased electrode diameter makes the gas gap of the DBD reactor shorter, which increases the reduced electric field E/N. As E/N increases, the mean electron energy increases as well; thus, increasing the inner electrode diameter makes it easier for the DBD reactor to generate O radicals and promote NO oxidation.

3. Conclusions

This Special Issue entitled "Plasma: From Materials to Emerging Technologies" covers only very few aspects of plasma applications. For readers this only touches on problems investigated by researchers specialized in plasma physics, chemistry and engineering. There are many scientific teams dealing with other plasma applications not presented here, such as biomedicine, agriculture, tar decomposition in process gas, water remediation soil remediation, food sterilization, hydrogen production, electrostatic precipitating, plasma-assisted combustion, electrohydrodynamics and many more. Fortunately, the Applied Sciences Editorial Board devoted and is going to devote several Special Issues to cover at least part of the research fields mentioned above.

Funding: This research received no external funding.

Acknowledgments: This Special Issues collected the efforts of all the authors, reviewers, and members of the Editorial Office of Applied Sciences. I would like to thank all the professional contributions to this publication.

Conflicts of Interest: The author declares no conflict of interest.

References

1. Martines, E.; Cavazzana, R.; Cordaro, L.; Zuin, M. The Helical Resonator: A Scheme for Radio Frequency Plasma Generation. *Appl. Sci.* **2021**, *11*, 7444. [CrossRef]
2. Luo, L.; Wang, Q.; Dai, D.; Zhang, Y.; Li, L. A Practical Method for Controlling the Asymmetric Mode of Atmospheric Dielectric Barrier Discharges. *Appl. Sci.* **2020**, *10*, 1341. [CrossRef]
3. Lee, M.; Ashikawa, N.; Jung, Y. Effect of Coulomb Focusing on the Electron—Atom Bremsstrahlung Cross Section for Tungsten and Iron in Nonthermal Lorentzian Plasmas. *Appl. Sci.* **2020**, *10*, 4832. [CrossRef]
4. Sobczyk, A.; Jaworek, A. Carbon Microstructures Synthesis in Low Temperature Plasma Generated by Microdischarges. *Appl. Sci.* **2021**, *11*, 5845. [CrossRef]
5. Li, D.; Tong, L.; Gao, B. Synthesis of Multiwalled Carbon Nanotubes on Stainless Steel by Atmospheric Pressure Microwave Plasma Chemical Vapor Deposition. *Appl. Sci.* **2020**, *10*, 4468. [CrossRef]
6. Krumpolec, R.; Jurmanová, J.; Zemánek, M.; Kelar, J.; Kováčik, D.; Černák, M. Selective Plasma Etching of Polymer-Metal Mesh Foil in Large-Area Hydrogen Atmospheric Pressure Plasma. *Appl. Sci.* **2020**, *10*, 7356. [CrossRef]
7. Cai, Y.; Lu, L.; Li, P. Study on the Effect of Structure Parameters on NO Oxidation in DBD Reactor under Oxygen-Enriched Condition. *Appl. Sci.* **2020**, *10*, 6766. [CrossRef]

applied sciences

Article

The Helical Resonator: A Scheme for Radio Frequency Plasma Generation

Emilio Martines *, Roberto Cavazzana, Luigi Cordaro and Matteo Zuin

Consorzio RFX, Corso Stati Uniti 4, 35127 Padova, Italy; roberto.cavazzana@igi.cnr.it (R.C.);
luigi.cordaro@igi.cnr.it (L.C.); matteo.zuin@igi.cnr.it (M.Z.)
* Correspondence: emilio.martines@igi.cnr.it

Abstract: The helical resonator is a scheme for the production of high voltage at radio frequency, useful for gas breakdown and plasma sustainment, which, through a proper design, enables avoiding the use of a matching network. In this work, we consider the treatment of the helical resonator, including a grounded shield, as a transmission line with a shorted end and an open one, the latter possibly connected to a capacitive load. The input voltage is applied to a tap point located near the shorted end. After deriving an expression for the velocity factor of the perturbations propagating along the line, and in the special case of the shield at infinity also of the characteristic impedance, we calculate the input impedance and the voltage amplification factor of the resonator as a function of the wave number. Focusing on the resonance condition, which maximizes the voltage amplification, we then discuss the effect of the tap point position, dissipation and the optional capacitive load, in terms of resonator performance and matching to the power supply.

Keywords: plasma; helical resonator; radio frequency; RF plasma source

check for **updates**

Citation: Martines, E.; Cavazzana, R.; Cordaro, L.; Zuin, M. The Helical Resonator: A Scheme for Radio Frequency Plasma Generation. *Appl. Sci.* **2021**, *11*, 7444. https://doi.org/10.3390/app11167444

Academic Editor: Mirosław Dors

Received: 8 May 2021
Accepted: 26 July 2021
Published: 13 August 2021

Publisher's Note: MDPI stays neutral with regard to jurisdictional claims in published maps and institutional affiliations.

1. Introduction

The generation in the laboratory of an ionized gas, also called plasma, requires, in most instances, application to the originally neutral gas of an electric field that is large enough to start an avalanche process of ionization events driven by free electrons (also called breakdown). While the required conditions greatly vary, depending on many variables related to the chosen experimental layout and working conditions, in broad terms, the typical applied voltages are in the range of a few hundreds volts to several kV. Furthermore, the voltage can be stationary or varying in time with different rates: the rate determines both the technology to be used to generate and transmit the required voltage and the physics of the breakdown process and of the produced plasma. Among the different possibilities, the radio frequency (RF) voltage is a possible choice. This term broadly represents voltages varying with a frequency ranging from around 1 MHz to several hundred MHz. RF plasmas are characterized by a lower voltage required for breakdown with respect to lower frequency devices, and by the possibility of avoiding contact between electrodes and plasma, especially when the inductive coupling regime is achieved.

In general terms, the production of a RF plasma requires a generator (possibly composed of an oscillator and an amplifier), and a matching network, which in this case, has the double role of matching the device input impedance to the output impedance of the generator, and to magnify the voltage to the required value for breakdown and subsequent plasma sustainment. The impedance matching feature is necessary to minimize the power reflected from the load to the generator. This is not an issue at lower frequencies, where the wavelength of the voltage disturbance is much larger than the apparatus size, so that instantaneous propagation can be assumed, and lumped element circuit treatments are appropriate. It becomes, however, important in the MHz range and beyond, where the wavelength becomes comparable to the system size. In this case, a finite time is required for voltage variation propagation, and transmission line theory comes into play. It is worth

5

noting that this brings many similarities between the field of plasma technology and that of radio transmission, although in the latter, voltage amplification is not a requirement, but rather an issue that should eventually be taken into account to avoid undesirable breakdown of air.

Many possibilities exist for matching network construction, and sophisticated solutions are now commercially available. There is, however, a solution that has already been used for plasma generation, but has not encountered great diffusion despite its potential simplicity: the helical resonator. The helical resonator, as will be described in more detail in the following, is an open-ended air core coil that can be used both for voltage amplification, exploiting a resonance condition, and for impedance matching, by proper design of its layout. The absence of a ferrite core implies that very low dissipation levels are attainable if proper care is put in the design of the device, allowing the use of relatively low-power generators.

It is worth making here the historical annotation that there is a strong connection between the helical resonator and the "extra coil" used by Nikola Tesla in his work at Colorado Springs to produce high voltage at high frequency [1]. Indeed, Tesla evolved from using a loosely coupled resonant transformer to a tightly coupled one, with the secondary circuit connected to an "extra coil", which was a capacitively loaded helical resonator. Keeping in mind that Tesla used frequencies of tens of kHz, so several issues related to dissipation are substantially different in our case (and the apparatus size as well), it is, however, important to remark that our present discussion is also useful for a better understanding of his work and of the modern day Tesla coil working principle.

Despite being a concept developed at the end of the XIX century, the literature on the helical resonator working principle is somehow scattered and also includes empirical formulas not derived from first principles [2]. This is also due to the fact that the same topic has been studied from several angles. Indeed, the concept is interesting for the following fields:

- Plasma generation [3–6];
- Ion trap antennas [7];
- Self-capacitance of coils [8,9];
- RF filters [10,11];
- Tesla coil construction [12];
- Precision measurements and metamaterial research [13–17].

In each of these fields, different authors treated the problem with different taste and emphasis on some issues, and collecting all these contributions in a unified view has proven to be not trivial. We thus would like to give a summary of the most important results on this subject, in the hope that this can prove useful to future scientists and practitioners who wish to properly understand this concept. In particular, we derive formulas that allow to predict the performance of a resonator from its basic properties, and illustrate a possible use for designing resonators, satisfying the given requirements. The whole discussion surrounds the resonator without plasma: the effects of a plasma formed inside the device will be the object of a forthcoming publication.

The paper is organized as follows: in Section 2, we introduce the geometry of the device we wish to describe, and the different flavors that it can take; in Section 3, we describe the propagation of electromagnetic perturbations in the resonator, deriving the main characteristics of the device, which are characteristic impedance and the velocity factor, as a function of its geometry. Armed with these results, we then move to Section 4 to model the resonator as a transmission line, looking for the resonance condition, which is the frequency that maximizes the voltage amplification at the open end, and for the input impedance, which should be adapted so as to match the output impedance of the generator. In Section 5, the special case of a resonator with the shield immediately near the helical conductor is treated, as this is both the most promising for plasma production and also the one that allows a fully analytical treatment. In Section 6, the effect of a capacitive load connected to the resonator (including parasitic capacitance) is discussed. The results found up to this point are then checked against the experimental data in Section 7. Finally, in Section 8, the conclusions are drawn.

2. Geometry

The basic geometry of a helical resonator, as defined in the context of this paper, is shown in Figure 1. The resonator consists of a helically wound conductor of length L, with one end (which we shall call bottom end in the following) grounded, while the other (top end) is left open, or connected to a load. We indicate with b the helix radius, with H its height, and with N the number of turns. The helix pitch is given by the following:

$$p = \frac{H}{N}. \tag{1}$$

The driving voltage, labeled V_{in}, is fed in an intermediate point, named "tap point", usually near the bottom end. The tap point axial position, normalized to H, is labeled as δ. The resonator is covered by a grounded conducting shield, of radius c. This shield, not present in all treatments (if absent, it may be considered at infinity) represents the second conductor of the transmission line for which the helix is the first one, according to the modeling presented in Section 4. The voltage at the top end of the resonator is labeled V_{out}.

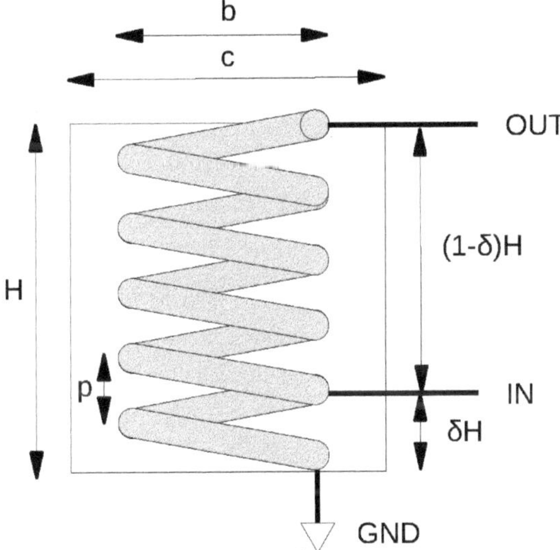

Figure 1. Schematics of the helical resonator. The terminals labeled IN and OUT are the ones where the voltages V_{in} and V_{out} are evaluated. The external rectangle represents the screen.

It is useful to emphasize that the shieldless resonator is fully defined by the three parameters (b, H, N), while in the more general case where the shield is present, the fourth parameter c is required. In principle, also the thickness of the helical conductor should be taken into account: however, this has only an effect on the dissipation due to the skin effect, and this effect is not so important at RF since the skin thickness is much smaller than any practical conductor thickness, so that the current is anyway flowing on the conductor surface. Therefore, we shall not consider this further parameter in our model, although one should be aware that p cannot be smaller than twice such thickness, and this puts a lower bound on the angle ψ defined below.

It is worthwhile to introduce the pitch angle ψ, defined as the following:

$$\tan \psi = \frac{p}{2\pi b}. \tag{2}$$

This will typically be quite small, expressing the fact that the helix is tightly wound, for the reason of compactness of the device; however, nothing prevents making it large, if required. The total length of the helical conductor can be expressed in terms of the other parameters as

$$L = N\sqrt{p^2 + 4\pi^2 b^2}. \tag{3}$$

In most practical situations, $p << 2\pi b$, so that $L \approx 2\pi b N$.

It is also useful to introduce the aspect ratio, defined as $H/(2b)$. This quantity gives an immediate idea of how elongated the coil is. In the following, we actually use, for convenience of notation, the inverse aspect ratio:

$$\epsilon = \left(\frac{H}{2b}\right)^{-1}. \tag{4}$$

It is useful for the following to notice that $\cot \psi = \pi N \epsilon$.

3. Propagation of Electromagnetic Perturbations into the Resonator

In order to properly model the helical resonator, we first need to understand how electromagnetic disturbances propagate in it, deriving a dispersion relation that is useful for obtaining the basic properties, such as phase velocity and, in a limit case, the characteristic impedance of the resonator seen as a transmission line. To accomplish this task, it is customary to model the resonator as a "sheath helix". This name is used to indicate an idealized anisotropically conducting cylindrical surface of infinite length, with infinite conductivity along the helix and zero conductivity normal to it. This model is possibly originally due to Ollendorf [18], and was treated by Sichak [19] and by Sensiper [20].

The starting point are Maxwell equations in vacuum as follows:

$$\nabla \times \mathbf{E} = -\frac{\partial \mathbf{B}}{\partial t} \tag{5}$$

$$\nabla \times \mathbf{B} = \frac{1}{c_0^2}\frac{\partial \mathbf{E}}{\partial t}. \tag{6}$$

Combining them, we obtain the wave equation as follows:

$$\frac{1}{c_0^2}\frac{\partial^2}{\partial t^2}\left\{\begin{matrix}\mathbf{E}\\\mathbf{B}\end{matrix}\right\} - \nabla^2\left\{\begin{matrix}\mathbf{E}\\\mathbf{B}\end{matrix}\right\} = 0 \tag{7}$$

where c_0 is the speed of light, which has as solutions in free space the electromagnetic plane waves.

We now consider the cylindrical geometry of the resonator and seek solutions of the form $\exp[i(\omega t - \beta z - m\theta)]$. Here, β is the axial wavevector, and m the azimuthal mode number. Furthermore, we restrict ourselves to $m = 0$ modes, because higher order modes are relevant only at very high frequencies. We thus deal with azimuthally symmetric perturbations of the form $\exp[i(\omega t - \beta z)]$. The wave equation becomes the Helmholtz equation as follows:

$$\left[\frac{1}{r}\frac{d}{dr}\left(r\frac{d}{dr}\right) - \beta^2 + \frac{\omega^2}{c_0^2}\right]\left\{\begin{matrix}\mathbf{E}\\\mathbf{B}\end{matrix}\right\} = 0. \tag{8}$$

Introducing $k = \omega/c_0$ (this is the wavevector of electromagnetic plane waves in vacuum), and defining the radial eigenvalue,

$$\tau^2 = \beta^2 - k^2, \tag{9}$$

we have the following:

$$\left(\frac{d^2}{dr^2} + \frac{1}{r}\frac{d}{dr} - \tau^2\right)\left\{\begin{matrix}\mathbf{E}\\\mathbf{B}\end{matrix}\right\} = 0. \tag{10}$$

The general solution of this equation is of the form $AI_0(\tau r) + BK_0(\tau r)$, where I_0 and K_0 are the modified Bessel functions of the first and second kind, respectively. It is important to remark, in order to avoid confusion, that k is here used as a normalized version of the frequency, whereas β is the actual wavenumber of the perturbations, and τ defines the radial oscillation of the electromagnetic field perturbation. Normally, $\beta \gg k$, so that $\tau \approx \beta$.

Let us now label with 1 the region inside the coil ($0 \leq r \leq b$) and with 2, the region between the coil and the shield ($b \leq r \leq c$). Taking into account that $K_0(x)$ diverges in $x = 0$, we have the following physically admissible solutions for the axial field components:

$$E_{z1} = A_1 I_0(\tau r) \tag{11}$$

$$B_{z1} = B_1 I_0(\tau r) \tag{12}$$

$$E_{z2} = A_2 I_0(\tau r) + A_2' K_0(\tau r) \tag{13}$$

$$B_{z2} = B_2 I_0(\tau r) + B_2' K_0(\tau r). \tag{14}$$

The other components of the fields are deduced from Maxwell equations as follows:

$$E_r = \frac{i\beta}{\tau^2} \frac{dE_z}{dr} \tag{15}$$

$$E_\theta = -\frac{i\omega}{\tau^2} \frac{dB_z}{dr} \tag{16}$$

$$B_r = \frac{i\beta}{\tau^2} \frac{dB_z}{dr} \tag{17}$$

$$B_\theta = \frac{i\omega}{c_0^2 \tau^2} \frac{dE_z}{dr} \tag{18}$$

We now apply the boundary conditions relevant for the sheath helix. On the helix surface ($r = b$) the longitudinal electric field has to be zero, due to the hypothesis of infinite conductivity. This gives the following:

$$E_{\theta 1}(b) \cos \psi + E_{z1}(b) \sin \psi = 0 \tag{19}$$

$$E_{\theta 2}(b) \cos \psi + E_{z2}(b) \sin \psi = 0. \tag{20}$$

The transverse component must be continuous, and since this is the only component present, this gives the following:

$$E_{\theta 1}(b) = E_{\theta 2}(b); \quad E_{z1}(b) = E_{z2}(b), \tag{21}$$

which makes redundant one of the two previous ones. The longitudinal component of the magnetic field must be continuous since there is no current flowing on the surface perpendicular to this direction, so that the following holds:

$$B_{\theta 1}(b) \cos \psi + B_{z1}(b) \sin \psi = B_{\theta 2}(b) \cos \psi + B_{z2}(b) \sin \psi. \tag{22}$$

Finally, on the conducting shield ($r = c$) the electric field will be zero, that is,

$$E_{\theta 2}(c) = 0; \quad E_{z2}(c) = 0. \tag{23}$$

which complete the set of six conditions required to fix the six constants A_1, B_1, A_2, A_2', B_2, B_2'.

After some algebra, the following eigenvalue equation for τ is obtained:

$$-(\tau b)^2 \frac{I_1(\tau c) I_0(\tau b)}{I_0(\tau c) I_1(\tau b)} \frac{(I_0(\tau c) K_0(\tau b) - I_0(\tau b) K_0(\tau c))}{(I_1(\tau b) K_1(\tau c) - I_1(\tau c) K_1(\tau b))} = (kb)^2 \cot^2 \psi. \tag{24}$$

This is the result that was obtained by Uhm and Choe [21], and also by Anicin [22], which is incorrectly reported (likely due to a mere transcription error) in Equation (25) of Niazi et al. [23]. It has the form $(\tau b)^2 g(\tau b, c/b) = (kb)^2 \cot^2 \psi$, which means that once the normalized frequency kb and the geometric factors c/b and ψ are specified, τb can be obtained, and subsequently, the normalized wave number βb using expression (9). The function $g(\tau b, c/b)$ is plotted in Figure 2 for different values of c/b. It can be seen that for large values of c/b, this is a strongly decreasing function of τb, whereas it tends to the constant value of 1 when the screen goes very near the coil. All the curves group together onto the unit value for large τb (indicatively, larger than 1). In Figure 2 are also reported the eigenvalues τb obtained from Equation (24), as a function of the parameter $kb \cot \psi$. The curves display an increasing behavior of the eigenvalue with the frequency, and converge to the relation $\tau b = kb \cot \psi$ as the shield approaches the coil.

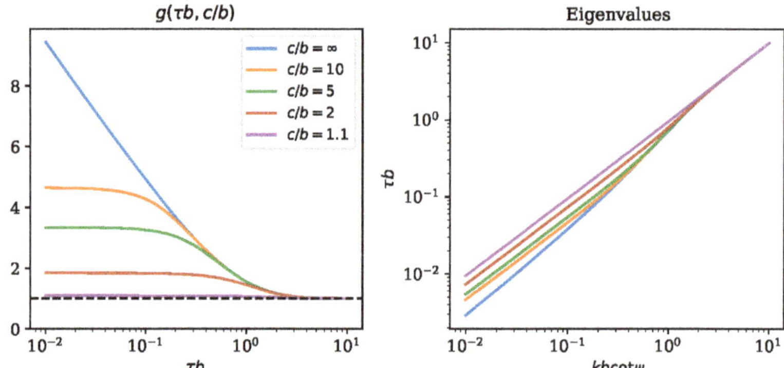

Figure 2. **Left**: function $g(\tau b, c/b)$ entering the eigenvalue equation, plotted as versus τb for different values of c/b. **Right**: solutions of the eigenvalue equation, plotted versus $kb \cot \psi$ for different values of c/b.

Once the eigenvalue equation is solved, one has the frequency as a function of the wave number, that is, the dispersion relation. It is then possible to evaluate the velocity factor, defined as the phase velocity v_p normalized to c_0:

$$V_f = \frac{v_p}{c_0} = \frac{1}{\sqrt{1 + \left(\frac{\tau b}{kb}\right)^2}}. \tag{25}$$

We should now recall that we are dealing with axial propagation, and therefore the velocity factor refers to the axial velocity. However, this holds in the sheath helix approximation, while in reality, the voltage pulse propagation takes place longitudinally, along the helically shaped conductor. If we consider the longitudinal propagaton along the helical conductor, the wave vector is given, through a simple projection, by $\beta^L = \beta \sin \psi$. Thus, the longitudinal velocity factor, given that the phase velocity is the ratio of the angular frequency to the wave number, is given by $V_f^L = V_f / \sin \psi$.

Such a longitudinal velocity factor is shown in Figure 3 for two different values of the pitch angle, corresponding to a tightly wound helix ($\psi = 1°$) and to a more loosely wound one ($\psi = 5°$). It can be seen that the longitudinal propagation takes place at a speed somehow larger than the speed of light in the vacuum, and that its frequency dependence is almost the same regardless of the pitch angle, with only a shift in the normalized frequency axis. We can identify three regions: at low frequency, the velocity factor is constant, with a value of 1 for the shield on the helix and increases to values larger than 3 when the shield is brought far away; a transition region; and a high frequency region where the propagation speed is equal to c_0, regardless of the shield position. We want to emphasize that the velocity factor is frequency dependent, so the propagation is dispersive.

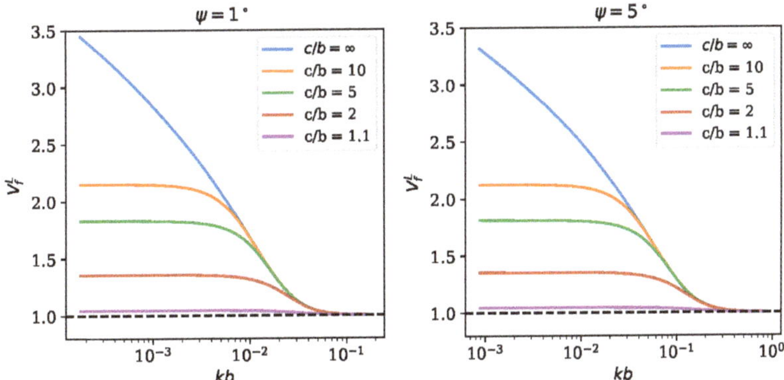

Figure 3. Longitudinal velocity factor V_f^L plotted as a function of the normalized frequency kb, for different values of the shield proximity c/b; the left panel refers to a pitch angle $\psi = 1°$, the right panel to $\psi = 5°$.

To put things in perspective, we can consider, for example, that for the industrial frequency $f = 27.12$ MHz, the plane wave wavelength is 11 m, so $k = 0.567$ m^{-1}. For a coil with a diameter of 10 cm, this gives $kb = 2.8 \times 10^{-2}$. This can be in the region where all longitudinal velocity factors collapse on a single curve or not, depending on the pitch angle value.

Before concluding this section, it is worth addressing in more detail two limit cases. The first one is the case without the shield, which, in our formulation, corresponds to $c/b \to \infty$. It is straightforward to show that in this case, the eigenvalue equation reduces to the following:

$$(\tau b)^2 \frac{I_0(\tau b)K_0(\tau b)}{I_1(\tau b)K_1(\tau b)} = (kb)^2 \cot^2 \psi. \qquad (26)$$

This result was quoted by several authors, in the context of analyzing the self-capacitance of a helical coil [24]. The rest of the analysis proceeds as before, but it is worth mentioning that the axial velocity factor can be approximated reasonably well, at least in the transition region, by the following expression [24]:

$$V_f \approx \frac{1}{\sqrt{1 + 20\left(\frac{d}{p}\right)^{2.5}\left(\frac{d}{\lambda_0}\right)^{0.5}}} \approx \frac{1}{\sqrt{1 + 0.645(\cot \psi)^{2.5}(kb)^{0.5}}}. \qquad (27)$$

where $d = 2b$ is the coil diameter. This is illustrated in Figure 4, which shows a superposition of the exact longitudinal velocity factor obtained from the resolution of the eigenvalue equation superposed to that obtained from the empirical formula above. It can be seen that, as long as the frequency is low enough so that the longitudinal velocity factor is larger than 1, the approximated value is reasonably close to the exact one.

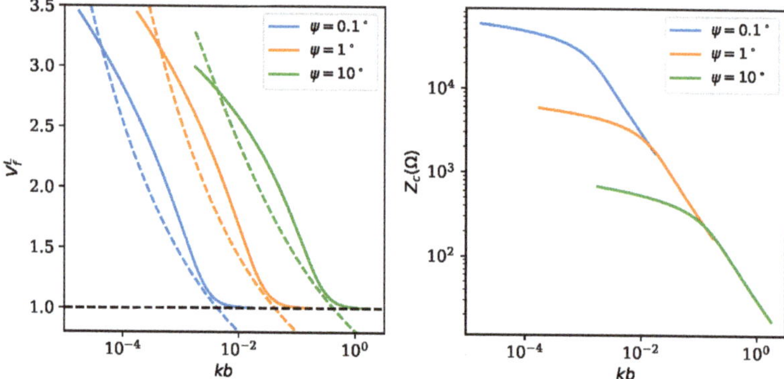

Figure 4. Left: exact longitudinal velocity factor V_f^L for the case without shield plotted as a function of the normalized frequency kb (solid lines) and approximation obtained from Equation (27) (dashed lines), for three different values of the pitch angle ψ. **Right**: characteristic impedance for the case without shield plotted as a function of the normalized frequency kb, for three different values of the pitch angle ψ.

For this particular case, an analytic formula for the characteristic impedance of the resonator seen as a transmission line was derived [24]. This is given by the following (expressed in Ohm):

$$Z_c = \frac{60}{V_f} I_0(\tau b) K_0(\tau b). \tag{28}$$

It is worth noting that the product of Bessel functions that appears in this formula is a decreasing function of its argument. The characteristic impedance for the case without shield is also plotted in Figure 4, as a function of the normalized frequency kb, for different pitch angles. It can be seen that the characteristic impedance decreases with frequency. For high frequencies, the curves all collapse on the following value:

$$Z_c^{HF} = \frac{30}{kb} \tag{29}$$

whereas at low frequencies, they fall below this curve. Typical characteristic impedance values for wavelengths much larger than the coil radius fall indicatively in the range of the $k\Omega$ for pitch angles between $1°$ and $10°$.

The second limit case is that in which the shield is directly over the coil, that is, $c/b = 1$. In this case, the eigenvalue equation simply reduces to the following:

$$\tau b = kb \cot \psi, \tag{30}$$

so that no numerical resolution is required. The dispersion relation becomes nondispersive.

$$k = \beta \sin \psi \tag{31}$$

with a velocity factor $V_f = \sin \psi$, and therefore a longitudinal velocity factor $V_f^L = 1$: the voltage pulses propagate along the conductor at the speed of light in vacuum.

We conclude this section by remarking that, in general, both the velocity factor and the characteristic impedance are functions of the frequency of the pitch angle ψ (which is the only coil geometrical parameter entering the treatment) and of the shield proximity c/b.

4. The Helical Resonator as a Transmission Line

Armed with the expressions for the velocity factor and, for the shieldless case , of the characteristic impedance, we are now in the position to model our helical resonator as a transmission line, with the helical coil as one conductor and the shield as the other, and with voltage V_{in} applied to the tap point. The basic concepts of transmission line modeling are recalled, for the reader's convenience, in Appendix A. The aim of this model is to obtain expressions for the two quantities, which are to be optimized for plasma generation, that is, the input impedance, which ideally should match the output impedance of the power supply (typically 50 Ω), and the voltage amplification factor that we want to maximize.

We start from basic Equations (A1) and (A2), describing the spatial behavior of voltage and current for a perturbation at frequency ω:

$$V(z) = V_0^+ e^{-\gamma z} + V_0^- e^{\gamma z} \tag{32}$$

$$I(z) = \frac{V_0^+}{Z_c} e^{-\gamma z} - \frac{V_0^-}{Z_c} e^{\gamma z} \tag{33}$$

where z is a coordinate running along the cylinder axis and $\gamma = \alpha + i\beta$, with α being the attenuation constant and β the wave number of the propagating perturbation. We consider separately the bottom and the top part of the resonator as two different transmission lines, each with its own boundary conditions. The bottom part, being short circuited ($V = 0$), gives the conditions at the grounded end ($z = 0$) and the tap point ($z = H\delta$):

$$V_0^+ + V_0^- = 0 \tag{34}$$

$$V_0^+ e^{-\gamma h} + V_0^- e^{\gamma h} = V_{in} \tag{35}$$

where we have introduced $h = H\delta$. These yield the following solution:

$$V_b(z) = V_{in} \frac{\sinh(\gamma z)}{\sinh(\gamma h)} \tag{36}$$

$$I_b(z) = \frac{V_{in}}{Z_c} \frac{\cosh(\gamma z)}{\sinh(\gamma h)}. \tag{37}$$

For the top part, which is open ($I = 0$) we have the following:

$$V_0^+ e^{-\gamma h} + V_0^- e^{\gamma h} = V_{in} \tag{38}$$

$$\frac{V_0^+}{Z_c} e^{-\gamma H} - \frac{V_0^-}{Z_c} e^{\gamma H} = 0 \tag{39}$$

yielding the following:

$$V_t(z) = V_{in} \frac{\cosh(\gamma(H - z))}{\cosh(\gamma(H - h))} \tag{40}$$

$$I_t(z) = \frac{V_{in}}{Z_c} \frac{\sinh(\gamma(H - z))}{\cosh(\gamma(H - h))}. \tag{41}$$

The voltage at the line endpoint is $V_{out} = V_t(H)$, so the following holds:

$$\frac{V_{out}}{V_{in}} = \frac{1}{\cosh(\gamma H(1 - \delta))}. \tag{42}$$

We shall call "voltage amplification factor" the modulus of this ratio. The total input current at the tap point is the following:

$$I_{in} = I_b(h) - I_t(h) = \frac{V_{in}}{Z_c} [\tanh(\gamma(H - h)) + \coth(\gamma h)] \tag{43}$$

so that the input impedance of the resonator is given by the following:

$$Z_{in} = \frac{V_{in}}{I_{in}} = \frac{Z_c}{\tanh(\gamma H(1-\delta)) + \coth(\gamma H\delta)}.$$

(44)

This is generally complex, with a resistive and a reactive part.

The expressions found for the input impedance and for the amplification factor are functions of the frequency, first because the wave number β is related to the frequency through the velocity factor ($k = \beta V_f$), and also because the velocity factor and the characteristic impedance are both frequency dependent. The dependence on the frequency of the attenuation coefficient α, which also exists, is neglected in the following.

In order to understand the behavior of the expressions found above, we first make the assumption that Z_c is independent of frequency, and we plot the relevant quantities as a function of the "electrical length" βH (which, we remind, is equal to $\beta^L L$), expressed in units of 2π. This quantity is related to the frequency through the velocity factor. The results are shown in Figure 5 for three different values of the attenuation factor αH.

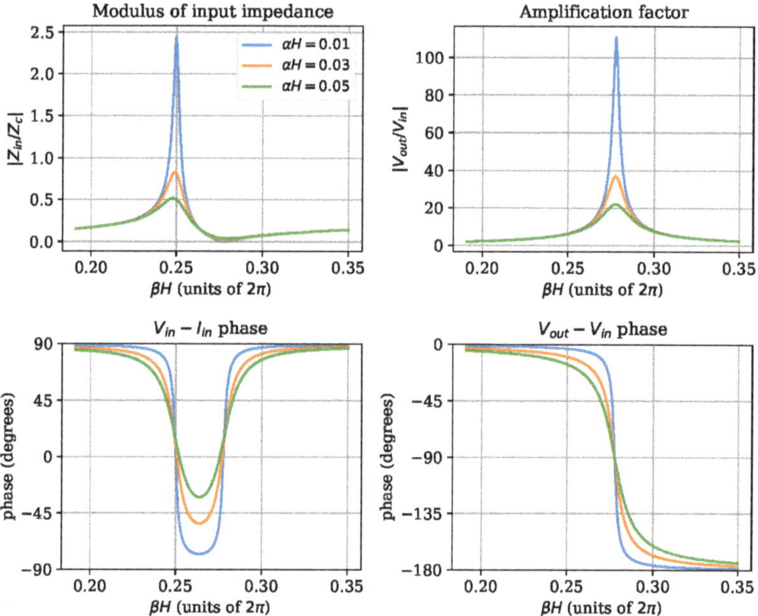

Figure 5. Behavior of the helical resonator modeled as a transmission line plotted as a function of the normalized wave number βH expressed in units of 2π, for three different values of the attenuation factor αH. The curves are computed for $\delta = 0.1$. Top left: modulus of the input impedance $|Z_{in}|$ normalized to the characteristic impedance Z_c. Top right: voltage amplification factor $|V_{out}/V_{in}|$. Bottom left: phase shift in degrees between input voltage and input current. Bottom right: phase shift in degrees between the input voltage and the output voltage.

The modulus of the input impedance displays a maximum and a minimum. The maximum is found for an electrical length as follows:

$$\beta_0 H = \frac{\pi}{2}.$$

(45)

This corresponds to the condition of the entire coil length equal to $\lambda/4$ (either considering the axial direction, with wavelength $2\pi/\beta$ and length H, or the longitudinal direction, with wavelength $2\pi/\beta^L$ and length L). It is worth noting that this point corresponds to a rather large impedance, of the order of the characteristic impedance, and thus in the range of the kΩ.

The voltage amplification factor displays a maximum, corresponding to the minimum of the impedance modulus. By analyzing expression (42), it can be seen that the maxima of the amplification factor occur at the following resonant wave numbers:

$$\beta_r H(1-\delta) = \frac{\pi}{2} + n\pi \qquad n = 0, 1, 2, \ldots. \tag{46}$$

which correspond to increasing resonant frequencies. These are $\lambda/4$ resonances computed by taking into account only the length of the top part of the resonator. The peak displayed in the graph, and the only one which will be considered in the following, is $\beta_r H(1-\delta) = \pi/2$. The extension of the results to higher order resonances is straightforward. Other authors have studied resonances of a higher order in the context of plasma production [25]. It is clear how the distance between the two critical points, that is, the point of maximum impedance and the resonance, increases as δ is increased. The amplification factor at the resonance is the following:

$$\left(\frac{V_{out}}{V_{in}}\right)_r = \frac{1}{i\sinh \alpha H(1-\delta)}. \tag{47}$$

Since in practical cases, $\alpha H \ll 1$, this can be approximated by the following:

$$\left(\frac{V_{out}}{V_{in}}\right)_r \approx \frac{1}{i\alpha H(1-\delta)}. \tag{48}$$

This expression elucidates the importance of achieving low dissipation in order to obtain a large voltage amplification.

Looking at the phases, it can be seen that the input voltage and current are in quadrature at low frequencies, then go in phase at the impedance maximum; then the phase is inverted, has another zero at the amplification factor maximum and returns to $\pi/2$ at high frequencies. Thus, the two critical points, that is, the maximum and minimum of the input impedance modulus, both approximately correspond to a real impedance. It is worth also noting that the extension of the phase reversal region between the two depends on the attenuation factor, so that this can be used as an indirect way of measuring it. Indeed, if the dissipation is too high, the phase will never become negative, and no purely resistive input impedance is possible. Finally, the input and output voltages are in phase at low frequencies, become in quadrature at the amplification factor maximum, as expected from Equation (48), and then go in phase opposition. As is shown in the next section, the distance between the two critical points depends on the parameter δ. In the following, we are concerned mainly with the point of maximum output voltage, which is related to a resonance of the system, and which represents the optimal condition for plasma generation. The voltage amplification becomes, of course, smaller as the attenuation factor increases, so minimizing dissipation is an important part of the design of an effective resonator.

A similar analysis can be performed by keeping the dissipation constant, and varying the tap point relative position δ. This is shown in Figure 6. It can be observed that the maximum modulus of impedance increases as δ is increased, while the amplification factor peak moves to the right, as expected, and grows in magnitude. The same shift occurs to the phase shift between the input and output voltage. The region of negative $V_{in} - I_{in}$ phase becomes deeper and wider. This last result shows that, for the given resonator parameters, there is a minimum δ that allows a purely resistive input impedance.

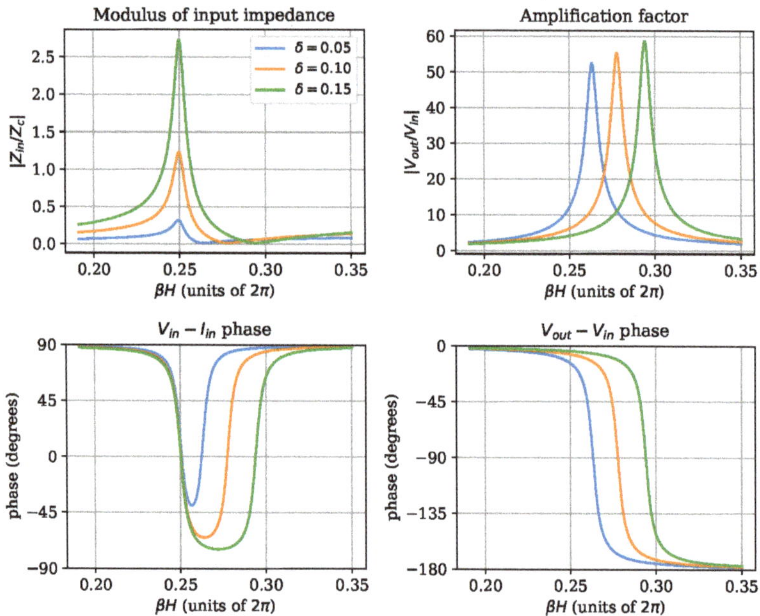

Figure 6. Behavior of the helical resonator modeled as a transmission line plotted as a function of the normalized wave number βH expressed in units of 2π, for three different values of the relative tap position δ. The curves are computed for $\alpha H = 0.02$. Top left: modulus of the input impedance $|Z_{in}|$ normalized to the characteristic impedance Z_c. Top right: voltage amplification factor $|V_{out}/V_{in}|$. Bottom left: phase shift in degrees between input voltage and input current. Bottom right: phase shift in degrees between the input voltage and the output voltage.

Since in the context of plasma production, the interest is focused on the resonance, which gives the maximum voltage amplification, we now plot the resistance at the peak, the resistance at the resonance, their ratio and the voltage amplification factor at the resonance, as a function of the dissipation factor αH for three δ values. This is shown in Figure 7. It can be seen that the peak resistance is of the same order of magnitude of the characteristic impedance, that is, of the order of the $k\Omega$, and decreases as the dissipation grows, while it increases when the tap point is moved to the right. On the contrary, the resistance at the resonance, which is the relevant parameter for the operation of the resonator as a voltage amplifier, increases both with dissipation and with tap position. This shows that changing the tap position is a way to achieve an impedance matching of the device with the power supply. Finally, the ratio of the output voltage to the input one is shown, as could be expected from energy considerations and from the previous figures, to decrease with dissipation, with a modest dependence on the tap position in the explored range. In the figure, we have also plotted the ratio of the peak resistance to the resistance at resonance, in order to illustrate the fact that this easily measurable parameter could be used as a way to estimate the attenuation factor αH. This, in turn, allows us to estimate the output without actually measuring it, a useful possibility given the fact that a direct measurement with a high voltage would be perturbative due to the addition of a capacitive load (as described below).

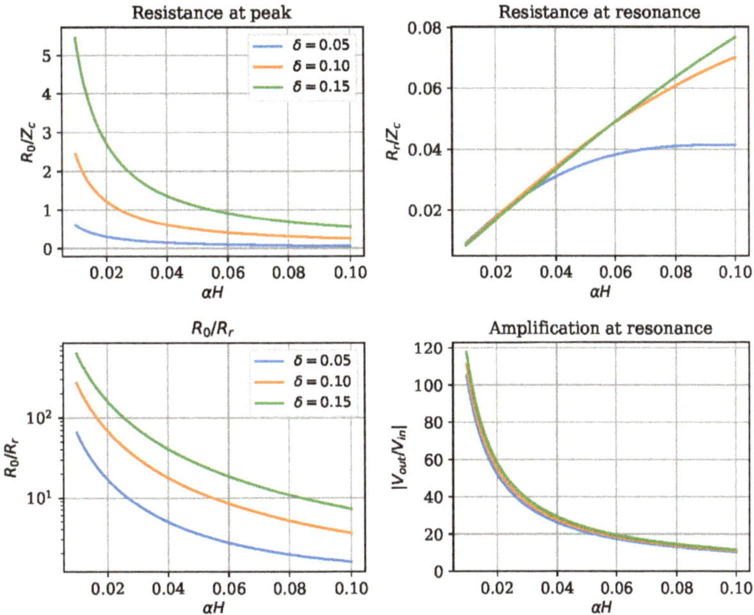

Figure 7. Behavior of the helical resonator at the peak resistance and at the resonance, plotted as a function of the normalized attenuation factor αH, for three different values of the relative tap position δ. Top left: input resistance at peak, normalized to the characteristic impedance Z_c. Top right: input resistance at resonance, normalized to the characteristic impedance Z_c. Bottom left: ratio of the previous two quantities. Bottom right: voltage amplification factor $|V_{out}/V_{in}|$.

We should now remark, though, that the input impedance at resonance is given by the following:

$$Z_{in}^r = \frac{Z_c}{\coth \alpha H (1 - \delta) + \coth \left[\alpha H \delta + i \frac{\pi}{2} \frac{\delta}{1 - \delta} \right]}. \tag{49}$$

This expression clarifies that the input impedance at resonance is not fully resistive. Indeed, the point of zero $V_{in} - I_{in}$ phase is located at a slightly lower frequency. As a consequence, the calculation of the input impedance and of the amplification factor for the point of purely resistive input impedance has to be done numerically by identifying the wave number for which the imaginary part of the impedance is zero, and then evaluating its real part at this same wave number. It is instructive to plot the resistance at the point of purely resistive impedance near resonance and the voltage amplification at the same point as a function of the tap position δ for different attenuation factor values. This is depicted in Figure 8. On the same figures, the resistance (real part of the impedance) and the amplification factor computed precisely at resonance are shown as dashed lines. It is possible to observe that the resistance curves at resonance display a steep rise for low values of δ, a maximum, and then a slower fall. This behavior suggests that if one wants to work at resonance and a specific value of this resistance is sought (typically 50 Ω for good matching), then one may wish to stay on the right of the peak, where the slow variation is forgiving in regards to imprecision in the construction. This also gives slightly higher voltage amplification than at very low δ values. Additionally, one should observe that if the dissipation is too low, then there exists the possibility that matching is not achievable because the input resistance is lower than that required for all possible δ values. However, if instead of working precisely at resonance, the nearby point of purely resistive input impedance is chosen, it can be seen that a monotonously decreasing resistance is found. In fact, for δ values that are sufficiently high, the curves superpose to those obtained at resonance, indicating a very small difference between the two points. A larger discrepancy appears at low δ, where the resistance at resonance decreases, while the other value increases steeply. In this same region, the amplification factor at the point of purely resistive impedance decreases with respect to the one at resonance. Generally speaking, it seems advisable to avoid the low δ values where a strong discrepancy of the two conditions appears, and where small errors in determining δ may translate to large changes in input resistance. For these values of the attenuation factor, this suggests a δ value of 0.1 or larger.

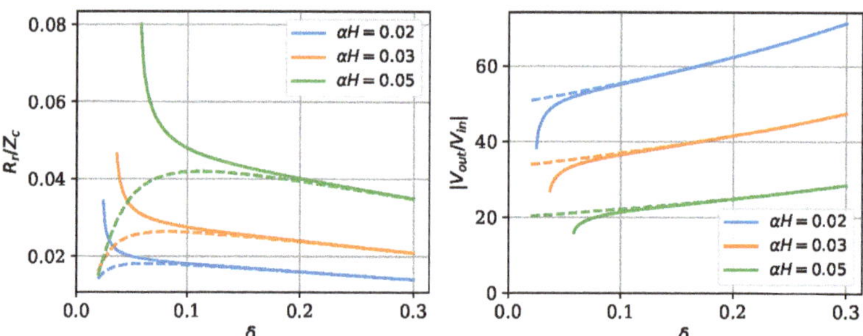

Figure 8. Behavior of the helical resonator at the point of purely resistive input impedance near resonance (continuous line) and at resonance (dashed line), plotted as a function of the tap position δ, for three different values of the normalized attenuation factor αH. **Left:** input resistance at resonance, normalized to the characteristic impedance Z_c. **Right:** voltage amplification factor $|V_{out}/V_{in}|$.

5. The Fully Shielded Helical Resonator

While the procedure described above constitutes the general approach, from now on, we want to focus on the condition $c/b = 1$, that is, with the shield directly superimposed to the helical coil. We shall call this case "fully shielded resonator". This is the most interesting situation for plasma production, because in practical situations, especially for devices that are intended to be taken out of the laboratory, one wishes to minimize electromagnetic radiation, both for safety and electromagnetic compatibility reasons. Since a helical antenna, when operating at wavelengths much larger than its size, radiates mainly from the sides [26], the grounded shield should prevent this irradiation (hence its name). Furthermore, since radiation is one major source of dissipation, it is expected that the shielded resonator will have a lower α. Thus, we think that a good construction practice is to always wrap the coil in a conducting grounded shield, with close proximity.

This situation makes also for easier calculations since, as seen above, no numerical solution is needed: $V_f = \sin \psi$ and $k = \beta \sin \psi$. This allows to recast in terms of the frequency the previously found expressions, which were given in terms of the wave number β. Indeed, it is straightforward to show that in this case, the following holds:

$$\beta H = \frac{2\pi N}{\cos \psi} kb. \tag{50}$$

The normalized frequency at which the peak in input resistance occurs is as follows:

$$k_0 b = \frac{\cos \psi}{4N} \approx \frac{1}{4N}. \tag{51}$$

We see that for a given coil diameter, this frequency is inversely proportional to the number of turns. It is also possible to normalize the frequency f to the frequency of maximum resistance f_0, and write the following:

$$\beta H = \frac{\pi}{2}\frac{f}{f_0}. \tag{52}$$

It is thus straightforward to replot the graphs in Figures 5 and 6 as a function of f/f_0. This is not the case for values of c/b larger than 1, where the velocity factor is frequency dependent, and the transformation from wave number to frequency is nonlinear. The resonance frequency f_r is related to f_0 by the following:

$$f_0 = f_r(1 - \delta). \tag{53}$$

6. Effect of a Capacitive Load

In many practical applications, the helical resonator will not be free-standing, but its upper end will be connected to some electrode or other structure, adding to it a load that, in a first approximation, can be considered fully capacitive. Even when this is not the case, one may wish to measure the voltage of the upper end with a high voltage oscilloscope probe, which will add a capacitive load of capacitance C (typically a few pF). We thus wish to address the problem of connecting a load with impedance

$$Z_L = \frac{1}{i\omega C} \tag{54}$$

to the transmission line model described above. Using a standard result of transmission line theory, the input impedance of a line of length l, and characteristic impedance Z_c connected to a load Z_L is the following:

$$Z_{in} = Z_c \frac{1 + \Gamma_L e^{-2\gamma H}}{1 - \Gamma_L e^{-2\gamma H}} \tag{55}$$

where Γ_L is the reflection coefficient at the load, given by the following:

$$\Gamma_L = \frac{Z_L - Z_c}{Z_L + Z_c}. \tag{56}$$

For the case of a capacitive load, the reflection coefficient takes the following form:

$$\Gamma_L = \frac{1 - i\omega\tau_L}{1 + i\omega\tau_L} \tag{57}$$

where the characteristic time τ_L is defined as follows:

$$\tau_L = Z_c C. \tag{58}$$

The helical resonator can be considered the parallel of the bottom section and of the top section, the latter being connected to the capacitive load. Since the bottom section is short-circuited by the ground connection, it has $Z_L = 0$ and, therefore, $\Gamma_L = -1$, so its input impedance is $Z_{in} = Z_c \tanh \gamma H\delta$. Combining this in parallel with the top section connected to the capacitive load, one obtains the overall input impedance as follows:

$$Z_{in} = Z_c \left[\coth \gamma H\delta + \frac{1 - \Gamma_L e^{-2\gamma H(1-\delta)}}{1 + \Gamma_L e^{-2\gamma H(1-\delta)}} \right]^{-1}. \tag{59}$$

In the limit $C \to 0$, Γ_L tends to 1, and the formula for Z_{in} correctly reduces to expression (44).

Using the fact that

$$\Gamma_L = \frac{V_0^- e^{\gamma H}}{V_0^+ e^{-\gamma H}} \tag{60}$$

it is straightforward to derive the voltage amplification factor:

$$\frac{V_{out}}{V_{in}} = \frac{1 + \Gamma_L}{e^{\gamma H(1-\delta)} + \Gamma_L e^{-\gamma H(1-\delta)}}. \tag{61}$$

According to Corum et al., this is "probably the most important equation in all of high voltage RF engineering" [27]. Once again, when Γ_L tends to 1, this formula reduces to expression (42).

Figure 9 shows, for a fully shielded resonator, the frequency dependence of the input impedance normalized to the characteristic impedance of the amplification factor, and of the phases between V_{in} and I_{in} and between V_{in} and V_{out}, for various values of $\omega_0 \tau_L$. As before, the frequency is normalized to the frequency corresponding to the wave number β_0, which satisfies $\beta_0 H = \pi/2$, that is, the frequency of the maximum input impedance in the no-load case.

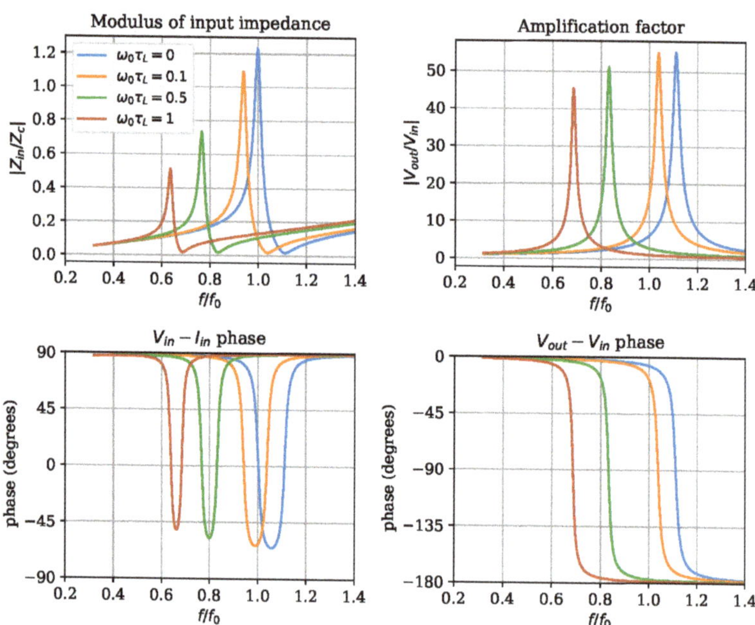

Figure 9. Behavior of the fully shielded helical resonator with capacitive load plotted as a function of frequency f, normalized to the frequency f_0 of maximum impedance modulus, for different values of $\omega_0 \tau_L$. The curves are computed for $\alpha H = 0.02$ and $\delta = 0.1$. Top left: modulus of the input impedance $|Z_{in}|$ normalized to the characteristic impedance Z_c. Top right: voltage amplification factor $|V_{out}/V_{in}|$. Bottom left: phase shift in degrees between input voltage and input current. Bottom right: phase shift in degrees between the input voltage and the output voltage.

It is possible to observe that as the load capacitance is increased (and, therefore, $\omega_0 \tau_L$ becomes larger) the resonance frequency decreases, and the peak impedance and peak voltage amplification are both reduced. The phase diagrams are also shifted toward lower frequencies. In general, the downshift in frequency appears to be the most relevant effect, unless the capacity becomes too large.

We can now evaluate as before the input impedance and the amplification factor at the near-resonance frequency where the input impedance is real. The results are shown in Figure 10, for four different values of $\omega_0 \tau_L$. We see that the curves have a shape similar to that shown in Figure 8, with a region with a strong gradient at low δ, difficult to practically use, and then a region of slowly varying Z_{in} and amplification factor. As the load capacitance is increased, the part with a steep gradient of the Z_{in} curves is shifted to the right, while the rest collapses onto a single curve, suggesting that the capacitive load does not affect the matching condition if one stays in this region. The amplification factor curves are shifted downward, indicating a reduction in voltage amplification caused by the increasing load. Since this gives a reduction in performance, care should be taken to avoid too large capacitance at the high voltage end of the resonator. If the resonator is to be connected to a system of electrodes, this should be taken into consideration.

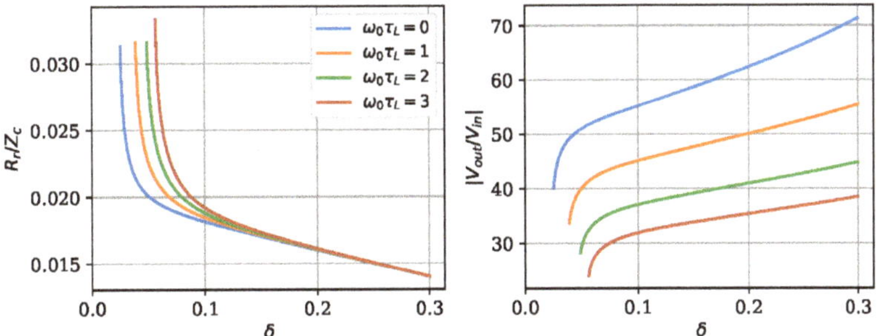

Figure 10. Left: plot of the input resistance normalized to the characteristic impedance of the fully shielded resonator with capacitive load, achieved in the purely resistive condition near the resonance, as a function of the relative tap point position δ, for different values of $\omega_0 \tau_L$. (**Right**): voltage amplification factor for the same conditions. The attenuation factor is $\alpha H = 0.02$.

7. Experimental Results

The theoretical results described up to now were checked against the properties of actual helical resonators, both with and without a conducting shield, in unloaded and loaded conditions. In these experiments, the input impedance was measured using a PocketVNA Vector Network Analyzer, in 1-port reflection only mode, which was properly calibrated. Furthermore, the voltage amplification factor at resonance was measured by sending RF power to the resonator and measuring the input voltage with a 10:1 oscilloscope probe and the output voltage with a 1000:1 Tektronix P6015a high voltage (HV) probe. For this latter experiment, the frequency was adjusted so as to achieve the maximum V_{out}/V_{in} ratio.

The first resonator (resonator 1) was built, using a 1 mm diameter insulated wire wound on a polyurethane tube of a 10 mm inner diameter and 12 mm outer diameter. Initially, 110 turns of wire were wound, and the tap point was located after 10 turns ($\delta = 0.091$). By measuring the resonator axial length, the average pitch was evaluated to be 1.04 mm, in good agreement with the wire thickness. Subsequently, while the tap point was kept fixed, the number of turns on the top part of the resonator was decreased from 100 to 90, then to 80, and so on up to 60 ($\delta = 0.143$).

The resonator input impedance modulus and phase were measured using the VNA as a function of frequency. The results are shown in the top row of Figure 11. The curves were fitted with expression (44). However, this expression is given as a function of the wavenumber β, which is related to the frequency by $k = \beta V_f$. Thus, in order to be able to perform a fit on the data given as a function of frequency, βH was replaced in expression (44) by $\pi f / 2 f_0$, introducing the new parameter f_0, which represents the frequency at which the electrical length of the entire resonator βH is equal to $\pi/2$, i.e., the frequency where the maximum of the input impedance is located. The fit thus led to optimal values of the parameters Z_c, αH and f_0, and then V_f was computed as $4Hf_0/c_0$. In fact, V_f is frequency dependent, and this dependence should be included in the fit. However, this would lead to a complex procedure, due to the need to solve the eigenvalue equation at each step. It was thus decided to assume V_f to be almost constant in the vicinity of the resonance. It can be seen that the resulting curves fit reasonably well with the experimental ones, both for the amplitude and the phase of the input impedance.

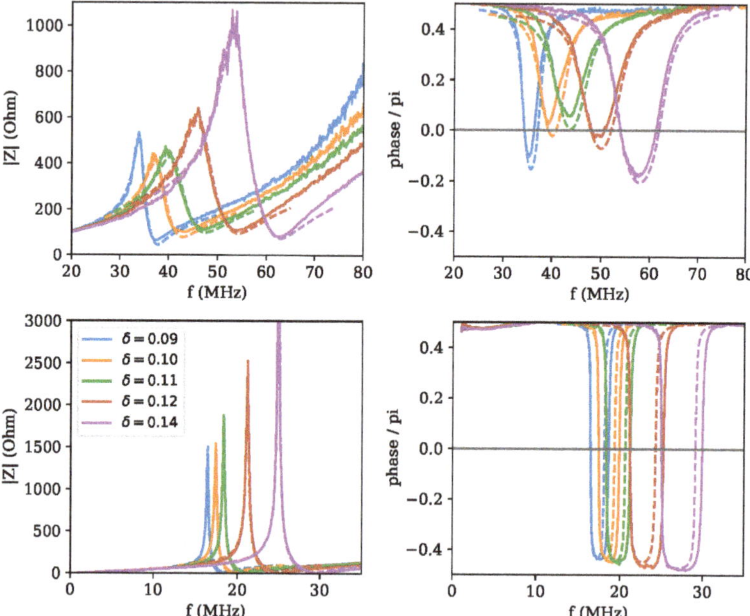

Figure 11. (**Top**): plot of the modulus (**left**) and phase (**right**), normalized to π, of the input impedance of resonator 1. The different curves correspond to different number of turns above the tap point. The dashed curves are the fits obtained as described in the text. **Bottom**: same as above, but with the addition of a conducting shield around the resonator.

Subsequently, a grounded shield made from an aluminum tube with an internal diameter of 16 mm and external diameter of 18.3 mm (average diameter of 17.15 mm and thickness of 2.3 mm) was applied to the resonator, and the fitting procedure was repeated as shown in the bottom row of Figure 11. The fitting curves also follow reasonably well the experimental ones. It can be noted that the curves for the shielded resonator are much sharper with stronger gradients, and the phases reach more negative values at the resonance, all indications of a reduced dissipation. This can be understood by assuming that the main source of dissipation is radiation from the sides of the device, which is prevented by the conducting screen. In this respect, it is important to notice, for experimenters wishing to repeat this work, that the wire used to ensure the grounding of the shield should be positioned so as to minimize the formation of loops, which can act as antennas.

The resulting values of the fit parameters, both without and with shield, are shown in Figure 12. It can be seen that the characteristic impedance without shield grows with δ from 1.15 kΩ to 1.3 kΩ, whereas with the shield, it has a constant value of around 580 Ω. The attenuation factor αH turns out to have a peaked shape, with values ranging between 0.05 and 0.11. This shape is somehow surprising: indeed, as δ is increased by reducing the number of turns on the top end, and therefore the resonator length H, one would expect a decreasing behavior of αH with δ, assuming α to be a constant independent of the resonator size. It is, however, possible that other issues enter the dissipation process, including possible different cable positions, and therefore different radiation patterns between one measurement and the following. This is also in agreement with the observation that in the shielded case, the obtained value of αH is constant, around 0.0085. If one assumes that the shield suppresses radiation, then this residual value could be attributed to other causes, and the case-to-case variation would disappear. The frequency f_0 was found to be linearly increasing with δ, both for the unshielded and shielded case. This was expected, in view of a constant velocity factor. Indeed, the velocity factor turns out to be essentially constant, with a value of 0.052 for the unshielded case and 0.024 for the shielded one.

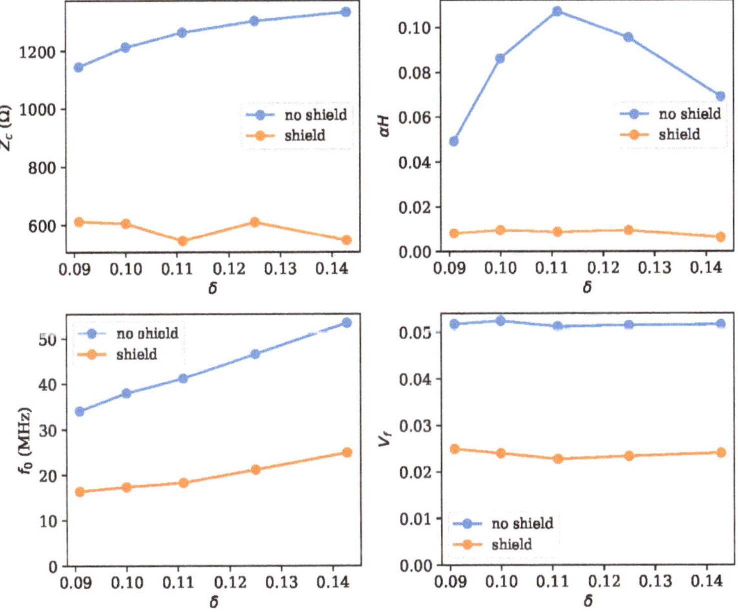

Figure 12. Parameters resulting from the fits of the input impedance data for resonator 1, plotted as a function of the relative distance δ of the tap point from the bottom end. Top left: characteristic impedance. Top right: attenuation parameter. Bottom left: frequency at which the total electrical length of the resonator is $\pi/2$. Bottom right: velocity factor, obtained from the previous quantity. In all graphs, results without and with conducting shield are reported.

An important issue to be addressed is how much the characteristic impedance (for the unshielded case) and the velocity factors match the predictions of the propagation model described in Section 3. According to expression (28), one would expect in the unshielded case a characteristic impedance of 2.25–2.45 kΩ at the frequency where the resonance occurs in the measurements. This is more or less double than what was found from the fits. Concerning the velocity factor, from expression (25) one would expect values of 0.07 for the unshielded case, and 0.036 for the shielded one. These are both 1.5 times larger than those found from the fits. The source of these discrepancies is not clear.

Subsequently, two other resonators were built (resonators 2 and 3) with a fixed number of turns (10 and 100 for the bottom and top parts, respectively), wound on quartz tubes. The tube of resonator 2 had a diameter of 12 mm, whereas the tube of resonator 3 had a diameter of 7 mm. The length H was respectively 110 mm and 103 mm. The shield was obtained from the same aluminum tube, which was used to shield resonator 1. Their input impedance was once again measured, using the Pocket VNA instrument. Since in this experiment, the aim was to measure the high voltage generated at the resonator open end, using a Tektronix P6015A high voltage probe, the input impedance measurements were performed both with and without the probe attached to the resonator. Indeed, the probe is expected to add a capacitive load, along the lines described in Section 6. The nominal HV probe input capacitance is, according to the technical specifications, around 3 pF.

The results of the measurements are shown in Figure 13 for resonator 2 and in Figure 14 for resonator 3. In both cases, the addition of the probe causes a downshift in frequency of the resonance pattern, both for the case without the shield (top row) and for the shielded case (bottom row). Again, the curves for the shielded case appear sharper, indication of a lower dissipation. The fitting procedure adopted is the following: at first, the curves without HV probe were fitted as described above. Subsequently, the curves obtained with the HV probe attached to the resonator open end were fitted, using expression (59), but keeping the values of Z_c and f_0 fixed to those resulting to the previous fit, and adjusting only αH and the new parameter τ_L. While, in principle one would think that also αH should be kept constant—and this was tried at the beginning—it was realized that it was not the case, possibly due to novel radiation sources stemming from the new layout.

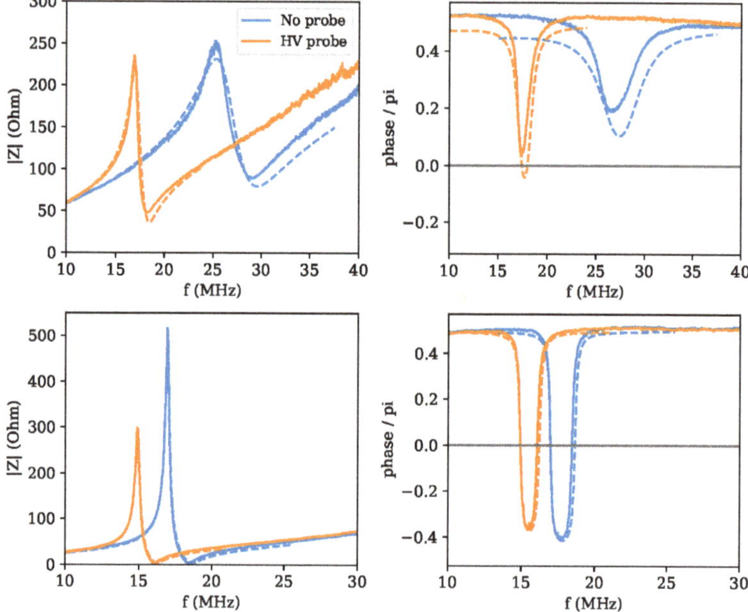

Figure 13. Top: plot of the modulus (**left**) and phase (**right**), normalized to π, of the input impedance of resonator 2, both in standard configuration and with the HV probe attached to the open end. The dashed curves are the fits obtained as described in the text. **Bottom**: same as above, but with the addition of a conducting shield around the resonator.

The results shown in Figure 13 for resonator 2 demonstrate that the model for the loaded resonator is actually very good. Indeed, the quality of the fit appears to be even better than for the resonator in standard conditions, suggesting that also in this case, some parasitic capacitance should be taken into account. Only the phases for the unshielded case show some discrepancy, mainly due to the experimental curves in the regions away from the resonance being somehow higher than the expected $\pi/2$ value. This is likely due to some instrumental error, which is not unlikely, given that the Pocket VNA is a low-cost instrument. Similar considerations apply also to the curves depicted in Figure 14, which refer to resonator 3.

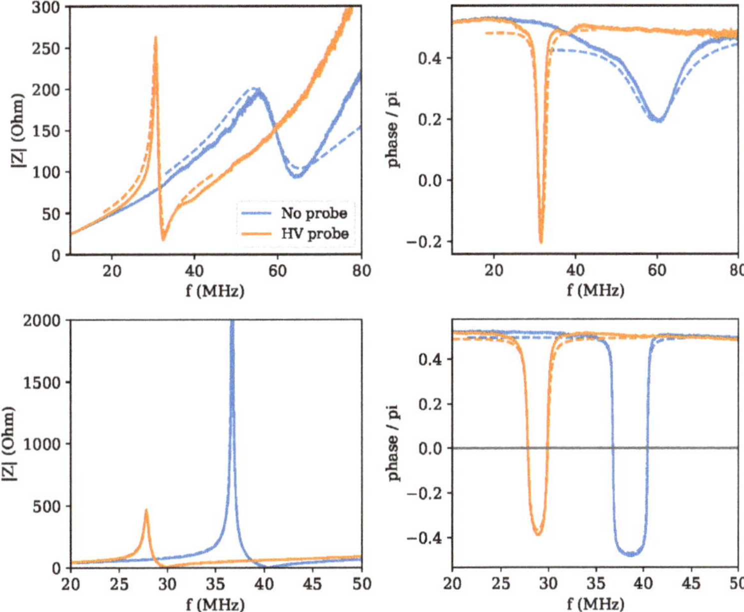

Figure 14. Top: plot of the modulus (**left**) and phase (**right**), normalized to π, of the input impedance of resonator 3, both in standard configuration and with the HV probe attached to the open end. The dashed curves are the fits obtained as described in the text. **Bottom**: same as above, but with the addition of a conducting shield around the resonator.

The results of the fits for the two resonators are summarized in Table 1. It can be seen that a characteristic impedance of about 1 kΩ was found for both resonators, which was then reduced to one quarter and to a half, respectively, when the shield was placed. The strong reduction in dissipation in the shielded case was confirmed. It is, however, to be noticed that also the placement of the HV probe on the high voltage end led to a reduction in dissipation, something which is not yet fully understood but which may depend on the radiation pattern. It is remarkable to observe that in the case of resonator 3, both measurements with the HV probe, with and without shield, led to a load capacitance of around 4 pF, not dissimilar to the nominal value of 3 pF. In the case of resonator 2, two slightly higher, and different, values of 5 and 7 pF were found. It is, however, to be noticed that the quality of contacts and other issues, such as the proximity of the work bench surface, may add parasitic capacitance. It is nevertheless remarkable that the correct order of magnitude could be obtained.

Concerning the comparison with the predictions of the propagation model of Section 3, for resonator 2 in the unshielded configuration, the model overestimates the characteristic impedance by a factor 2.8 and the velocity factor by a factor 2.6, while in the shielded case, the overestimate is by a factor 1.9. For resonator 3 in the unshielded configuration, there is an overestimation of the characteristic impedance by a factor of 1.6 and of the velocity factor by a factor of 2.2, whereas in the shielded case, the velocity factor is overestimated by a factor 1.9. Overall, these results confirm that the propagation model needs to be somehow updated to better match the experimental results.

Table 1. Parameters resulting from the fit of the input impedance curves for resonators 2 and 3. The load characteristic time τ_L and the associated load capacitance C are also reported.

	Shield	HV Probe	Z_c (Ω)	αH	V_f	τ_L (ns)	C (pF)
Resonator 2	No	No	1003	0.119	0.025	-	-
Resonator 2	No	Yes	1003	0.047	0.025	5.58	5.56
Resonator 2	Yes	No	272.4	0.011	0.016	-	-
Resonator 2	Yes	Yes	272.4	0.015	0.016	2.04	7.48
Resonator 3	No	No	1064	0.173	0.055	-	-
Resonator 3	No	Yes	1064	0.026	0.055	4.54	4.26
Resonator 3	Yes	No	501.7	0.004	0.036	-	-
Resonator 3	Yes	Yes	501.7	0.013	0.036	2.29	4.57

The voltage amplification factor was measured for both resonators 2 and 3 by tuning the frequency on the resonance, that is, on the condition of maximum amplification. The output voltage was measured, using the Tektronix P6015A high voltage probe, whereas the input voltage was measured with a standard x10 oscilloscope probe, which was tested and shown not to alter significantly the circuit behavior. The resulting voltage amplification factors, for both resonators 2 and 3, unshielded and shielded, are given in Table 2, where they are compared with the predicted values given by $1/\alpha H$, with αH resulting from the fits. It can be seen that a reasonable agreement is achieved, confirming that the modeling described above is adequate to predict the performance of a given resonator. It should be emphasized that, as expected from the fit results, the best performance is given by the shielded resonators. In particular, resonator 3 achieved a voltage amplification of 100, while resonator 2 achieved a value of 80. In this latter case, this resulted in an actual measured V_{out} of 4.3 kV—more than appropriate to ionize most gases in a wide pressure range. It is to be remarked that the actual output voltage will depend on the input one, which in itself will be determined by the power of the amplifier used as the power source, and by the input impedance.

Table 2. Comparison of the experimental amplification factor at resonance with the expected value $1/\alpha H$ predicted from the fit outcome, for resonators 2 and 3, with and without shield.

	Shield	$1/\alpha H$	V_{out}/V_{in}
Resonator 2	No	21.4	41
Resonator 2	Yes	61.8	80
Resonator 3	No	38.0	30
Resonator 3	Yes	77.6	100

8. Conclusions

The production of RF plasmas requires both a good matching of the load to the power supply and a method to magnify the voltage so as to achieve the gas breakdown. The shielded helical resonator excited at a tap point near the grounded end is a concept that allows, in principle, to achieve both goals, allowing a very simple construction for the plasma source and a direct connection to the generator, without the need for a fault-prone matching network. However, in order to achieve both goals, a proper design is needed. In this paper, we derived formulas that allow a prediction of the resonator performance, thus enabling the plasma scientist to properly design his source according to the parameters required for his plasma, and we have tested them experimentally.

Clearly, what is missing in this treatment is the effect of the plasma itself, once it is ignited. This puts an additional load on the resonator, and it is to be evaluated how much it can bring the operational conditions away from the optimal ones, especially in relation to the reflected power. This will be the topic of a subsequent study.

Similarly, the results concerning the treatment of the resonator as a transmission line depend on the characteristic impedance, which, for the moment, can be explicitly calculated only for the shieldless case and needs to be measured experimentally in all other situations, particularly for the fully shielded resonator. Furthermore, even in the shieldless case, it was found that the formula gave a result different from the experimental one by a factor of two, something which needs to be better investigated.

It is also to be remarked that to date, we have implicitly considered a capacitively coupled plasma source, where the amplified voltage gives rise to an electrostatic field, inducing breakdown. However, the very nature of the helical resonator makes it suitable also for the transition to an inductive regime [28]. This possibility and the effects on the load need also to be investigated.

In conclusion, it is our belief that a renewed interest in the helical resonator concept could lead to new, simpler and more compact designs for RF plasma sources, increasing the efficiency and the reliability. Furthermore, the concept of a circuit element where the lumped circuit approximation fails and the pattern and speed of voltage propagation become important, is in itself a stimulating and non-trivial idea that should be more widely taught in plasma technology courses.

Author Contributions: Conceptualization, E.M., R.C., L.C. and M.Z.; methodology, E.M.; software, E.M.; validation, R.C., L.C. and M.Z.; writing—original draft, E.M.; writing—review and editing, E.M., R.C., L.C. and M.Z.; visualization, E.M.; supervision, E.M. All authors have read and agreed to the published version of the manuscript.

Funding: This research received no external funding.

Conflicts of Interest: The authors declare no conflict of interest.

Abbreviations

The following abbreviations are used in this manuscript:

RF   Radio Frequency
HV   High Voltage

Appendix A. Transmission Line Theory

We recall here the basic concepts of transmission line theory. A transmission line is characterized by four distributed line parameters: the resistance per unit length R, the inductance per unit length L, the capacitance per unit length C and the conductance per unit length G of the dielectric separating the two conductors of the line.

The general solution at an angular frequency ω for the voltage and current along a line, composed of the superposition of a forward-propagating wave and a backward-propagating one is the following:

$$V(x) = V_0^+ e^{-\gamma x} + V_0^- e^{\gamma x} \tag{A1}$$

$$I(x) = \frac{V_0^+}{Z_c} e^{-\gamma x} - \frac{V_0^-}{Z_c} e^{\gamma x} \tag{A2}$$

where x is a coordinate running along the line, $\gamma = \alpha + i\beta$ with α being the attenuation constant and β the wave number of the propagating perturbations, and Z_c is the characteristic impedance of the line. These quantities are defined by the distributed line parameters, in the limit $R \ll \omega L$ and $G \ll \omega C$, by the following:

$$\alpha = \frac{1}{2}\left(\frac{R}{Z_c} + GZ_c\right) \qquad \beta = \frac{\omega}{v_p} \tag{A3}$$

where the characteristic impedance and the phase velocity are given by the following:

$$Z_c = \sqrt{\frac{L}{C}} \qquad v_p = \frac{1}{\sqrt{LC}}. \tag{A4}$$

It is worth noting that, if G is negligible,

$$\alpha = \frac{R}{2Z_c}. \tag{A5}$$

The actual values of the distributed parameters are difficult to predict from the specifications of the line for non-trivial geometries, so it seems better to consider v_p (or the velocity factor $V_f = v_p/c_0$), Z_c and α as unknown parameters and derive them either by studying the propagation properties of the line or by the fitting of the experimental data. The attenuation coefficient α, however, remains still undetermined, and cannot be trivially computed from Equation (A5) because the resistance per unit length can be substantially different from the DC value and must be evaluated experimentally.

The reflection coefficient of the line is a position-dependent quantity defined as follows:

$$\Gamma(x) = \frac{V_0^- e^{\gamma x}}{V_0^+ e^{-\gamma x}} = \frac{V_0^-}{V_0^+} e^{2\gamma x} = \Gamma(0) e^{2\gamma x} \tag{A6}$$

The wave impedance is another position-dependent quantity defined as follows:

$$Z(x) = \frac{V(x)}{I(x)} = Z_c \frac{V_0^+ e^{-\gamma x} + V_0^- e^{\gamma x}}{V_0^+ e^{-\gamma x} - V_0^- e^{\gamma x}} = Z_c \frac{1 + V_0^-/V_0^+ e^{2\gamma x}}{1 - V_0^-/V_0^+ e^{2\gamma x}}. \tag{A7}$$

Recalling the definition of the reflection coefficient, we have the following:

$$Z(x) = Z_c \frac{1 + \Gamma(0) e^{2\gamma x}}{1 - \Gamma(0) e^{2\gamma x}} = Z_c \frac{1 + \Gamma(x)}{1 - \Gamma(x)}. \tag{A8}$$

The reflection coefficient can be written in terms of impedance as follows:

$$\Gamma(x) = \frac{Z(x) - Z_c}{Z(x) + Z_c}. \tag{A9}$$

Voltage and current can be written in terms of the reflection coefficient as follows:

$$V(x) = V_0^+ e^{-\gamma x}(1 + \Gamma(x)) \tag{A10}$$

$$I(x) = \frac{V_0^+}{Z_c} e^{-\gamma x}(1 - \Gamma(x)). \tag{A11}$$

References

1. Cvetic, J.M. Tesla's high voltage and high frequency generators with oscillatory circuits. *Serbian J. Electr. Eng.* **2016**, *13*, 301–333. [CrossRef]
2. Macalpine, W.W.; Schildknecht, R.O. Coaxial resonators with helical inner conductor. *Proc. IRE* **1959**, *47*, 2099–2105. [CrossRef]
3. Niazi, K.; Lieberman, M.A.; Lichtenberg, A.J.; Flamm, D.L. Operation of a helical resonator plasma source. *Plasma Sources Sci. Technol.* **1994**, *3*, 452–465. [CrossRef]
4. Welton, R.F.; Thomas, E.W.; Feeney, R.K.; Moran, T.F. Simple method to calculate the operating frequency of a helical resonator-RF discharge tube configuration. *Meas. Sci. Technol.* **1991**, *2*, 242–246. [CrossRef]
5. Hopwood, J. Review of inductively coupled plasma for plasma processing. *Plasma Sources Sci. Technol.* **1992**, *1*, 109–116. [CrossRef]
6. Bongkoo, K.; Park, J.C.; Kim, Y.H. Plasma Uniformity of Inductively Coupled Plasma Reactor with Helical Heating Coil. *IEEE Trans. Plasma Sci.* **2001**, *29*, 383–387. [CrossRef]
7. Siverns, J.D.; Simkins, L.R.; Weidt, S.; Hensinger, W.K. On the application of radio frequency voltages to ion traps via helical resonators. *Appl. Phys. B* **2012**, *107*, 921–934. [CrossRef]
8. Medhurst, R.G.H.F. resistance and self-capacitance of single layer solenoids. *Wirel. Eng.* **1947**, *24*, 35–43.
9. De Miranda, C.M.; Pichorim, S.F. Self-resonant frequencies of air-core single-layer solenoid coils calculated by a simple method. *Electr. Eng.* **2015**, *97*, 57–64. [CrossRef]
10. Everard, J.K.A.; Cheng, K.K.M.; Dallas, P.A. High-Q helical resonator for oscillators and filters in mobile communication systems. *Electron. Lett.* **1989**, *25*, 1648–1650. [CrossRef]

11. Antoniuk, J.; Żukociński, M.; Abramowicz, A.; Gwarek, W. Investigation of Resonant Frequencies of Helical Resonators. In Proceedings of the 11th Microcoll, Budapest, Hungary, 10–11 September 2003; p. 5.
12. Blažević, Z.; Poljak, D. Some notes on transmission line representations of Tesla's transmitters. In Proceedings of the 16th International Conference on Software, Telecommunications and Computer Networks, Split, Croatia, 25–27 September 2008; pp. 60–69.
13. Bletzinger, P. Dual mode operation of a helical resonator discharge. *Rev. Sci. Instrum.* **1994**, *65*, 2975. [CrossRef]
14. Deng, K.; Sun, Y.L.; Yuan, W.H.; Xu, Z.T.; Zhang, J.; Lu, Z.H.; Luo, J. A Modified Model of Helical Resonator with Predictable Loaded Resonant Frequency and Q-Factor. *Rev. Sci. Instrum.* **2014**, *85*, 104706. [CrossRef] [PubMed]
15. Deri, R.J. Dielectric measurements with helical resonators. *Rev. Sci. Instrum.* **1986**, *57*, 82. [CrossRef]
16. Yang, J.H.; Zhang, Y.; Li, X.M.; Li, L. An improved helical resonator design for rubidium atomic frequency standards. *IEEE Trans. Instrum. Meas.* **2010**, *59*, 1678. [CrossRef]
17. Zhu, J.W.; Hao, T.; Stevens, C.J.; Edwards, D.J. Optimal design of miniaturized thin-film helical resonators. *Appl. Phys. Lett.* **2008**, *93*, 234105. [CrossRef]
18. Ollendorf, F. *Die Grundlagen der Hochfrequenztechnik*; Springer: Berlin, Germany, 1926; pp. 79–87.
19. Sichak, W. Coaxial Line with Helical Inner Conductor. *Proc. IRE* **1954**, *42*, 1315–1319. [CrossRef]
20. Sensiper, S. Electromagnetic Wave Propagation on Helical Structures (A Review and Survey of Recent Progress). *Proc. IRE* **1955**, *43*, 149–161. [CrossRef]
21. Uhm, H.S.; Choe, J.-Y. Properties of the electromagnetic wave propagation in a helix-loaded waveguide. *J. Appl. Phys.* **1982**, *53*, 8483. [CrossRef]
22. Anicin, B.A. Plasma loaded helical waveguide. *J. Phys. D Appl. Phys.* **2000**, *33*, 1276–1281. [CrossRef]
23. Niazi, K.; Lichtenberg, A.J.; Lieberman, M.A. The dispersion and matching characteristics of the helical resonator plasma source. *IEEE Trans. Plasma Sci.* **1995**, *23*, 833. [CrossRef]
24. Corum, K.L.; Corum, J.F. RF coils, helical resonators and voltage magnification by coherent spatial modes. In Proceedings of the 5th International Conference on Telecommunications in Modern Satellite, Cable and Broadcasting Service. TELSIKS 2001. Proceedings of Papers (Cat. No.01EX517), Nis, Yugoslavia, 19–21 September 2001; Volume 1, p. 339.
25. Park, J.-C.; Kang, B. Impedance model of helical resonator discharge. *IEEE Trans. Plasma Sci.* **1997**, *25*, 1398–1405. [CrossRef]
26. Kraus, J.D. The helical antenna. *Proc. IRE* **1949**, *37*, 263–272. [CrossRef]
27. Corum, J.; Daum, J.; Moore, H.L. *Tesla Coil Research*; Contractor Report ARCCD.CR-92006; U.S. Army Armament Research, Development and Engineering Center: Picatinny, NJ, USA, 1992.
28. Park, J.-C.; Lee, J.K.; Kang, B. Properties of inductively coupled plasma source with helical coil. *IEEE Trans. Plasma Sci.* **2000**, *28*, 403–413. [CrossRef]

Article

Carbon Microstructures Synthesis in Low Temperature Plasma Generated by Microdischarges

Arkadiusz T. Sobczyk * and Anatol Jaworek *

Institute of Fluid Flow Machinery, Polish Academy of Sciences, 80-231 Gdansk, Poland
* Correspondence: sobczyk@imp.gda.pl (A.T.S.); jaworek@imp.gda.pl (A.J.)

Abstract: The aim of this paper is to investigate the process of growth of different carbon deposits in low-current electrical microdischarges in argon with an admixture of cyclohexane as the carbon feedstock. The method of synthesis of carbon structures is based on the decomposition of hydrocarbons in low-temperature plasma generated by an electrical discharge in gas at atmospheric pressure. The following various types of microdischarges generated at this pressure were tested for both polarities of supply voltage with regard to their applications to different carbon deposit synthesis: Townsend discharge, pre-breakdown streamers, breakdown streamers and glow discharge. In these investigations the discharge was generated between a stainless-steel needle and a plate made of a nickel alloy, by electrode distances varying between 1 and 15 mm. The effect of distance between the electrodes, discharge current and hydrocarbon concentration on the obtained carbon structures was investigated. Carbon nanowalls and carbon microfibers were obtained in these discharges.

Keywords: plasma; microdischarge; electrical discharge; dusty plasma; hydrocarbon; carbon structures

Citation: Sobczyk, A.T.; Jaworek, A. Carbon Microstructures Synthesis in Low Temperature Plasma Generated by Microdischarges. *Appl. Sci.* **2021**, *11*, 5845. https://doi.org/10.3390/app11135845

Academic Editor: Emilio Martines

Received: 30 April 2021
Accepted: 16 June 2021
Published: 23 June 2021

Publisher's Note: MDPI stays neutral with regard to jurisdictional claims in published maps and institutional affiliations.

1. Introduction

Methods of production and functionalization of different carbon structures has been the subject of intense investigations in recent years. It has passed less than 30 years since Iijima published his results of experimental investigations on the production of carbon nanotubes in a high-current arc discharge. Since that time, many papers have been published in the field of the synthesis of carbon nanotubes, carbon onions, diamond-like carbon using various types of electrical discharges, chemical vapor deposition or laser ablation.

Plasma generated by electrical discharges is used in various industrial processes, for example, for thin film deposition, surface properties modification, sterilization, noxious compound decomposition, particulate matter precipitation, micro- and nanostructures synthesis, etching or electric-discharge-machining. These processes, operating at pressures ranging from the atmospheric to the elevated to below one pascal, are generated by a high voltage of various frequencies, such as direct-current (DC), alternating-current (AC), radio-frequency (RF), microwaves (μW) or by pulsed discharges (PD). These discharges can be generated between the following electrodes of different geometries: needle-plate, two parallel plates, rod-plate, etc., with or without dielectric barrier between them. Nowadays, a high-current arc discharge is one of the most effective methods used to obtain single- and multi-walled carbon nanotubes [1]. The synthesis of carbon structures or thin film deposition in low-power electrical discharges is rarely met in the literature. High voltage, low current discharges, such as corona discharge, generate non-equilibrium plasma, but can produce electrons of high kinetic energy, advantageous for such processes as ionization, excitation, molecules decomposition or polymerization.

Usually, microplasma is generated in a gas at or near atmospheric pressure (0.1 MPa) between electrodes spaced at small distances, ranging from a fraction of a millimeter to 10 mm. The word "microplasma" was probably used for the first time by Tachibana in 2003 [2], but "microdischarge" was found in a paper by Martynov and Ivanov [3] (see [4]).

Appl. Sci. **2021**, *11*, 5845. https://doi.org/10.3390/app11135845

https://www.mdpi.com/journal/applsci

The main advantage of using microplasma-based processes is that the physical processes and chemical reactions can occur in high-pressure gas flowing continuously through a plasma reactor, contrary to low pressure plasma sources, which require vacuum installation. In materials processing science, microplasma is used to synthesize nanoscale structures such as, for example, SiO_2 thin films [5], Si nanocones [6], carbon dendrites [7], or carbon amorphous deposits [8]. Generally, the most popular discharge types for microplasma generation are plasma-jet discharge and dielectric barrier discharge.

The investigations of carbon fiber synthesis between a needle and plate electrodes were presented by Brock and Lim [9], but in n-heptane vapors, and nitrogen as a carrier gas. The source of plasma was a corona glow discharge of negative polarity (a needle was used as the cathode). Carbon fibers of grained and irregular morphology were obtained as a result of discharge in that gas mixture. Similar structures, named carbon dendrites, were obtained by Kozak et al. using a glow discharge generated in ethanol as the carbon feedstock [7]. Raman spectra indicated that the synthesized dendrites have glassy carbon- and pyrocarbon-like structures. A negative corona discharge was also used to deposit carbon dendrites from the products obtained by the de-polymerization of polystyrene [10].

In recent years, low-current electrical discharges have also been tested as a method of carbon nanostructures synthesis. For example, Li et al. [11] synthesized carbon nanotubes in a corona discharge at atmospheric pressure. During electrical discharge in a methane/hydrogen mixture using an anodic aluminum oxide template on a stainless-steel plate and cobalt as a catalyst, multi-walled carbon nanotubes with a diameter of about 40 nm were obtained. The discharge was generated from a tungsten needle supplied by an AC voltage source of 8 kV/25 kHz. Sano and Nobuzawa [12] also obtained carbon nanotubes (CNTs) using a tungsten needle as the DC discharge electrode, but without a catalyst. A flat-ended graphite rod was used as a ground electrode and a tungsten wire as the cathode. The feedstock of carbon was ethylene mixed with hydrogen. (C_2H_4:H_2 (1:100)). The discharge voltage and current were 1.6 kV and 0.3 mA, respectively. The other structures obtained, using a negative corona discharge, were carbon nanowalls. During glow discharge with a rod used as the cathode and a plate as the anode, Mesko et al. [13] synthesized carbon nanowalls from the vapors of ethanol or hexane. Sobczyk, using cyclohexane as the carbon feedstock in a positive corona discharge, obtained carbon fibers synthesized at the needle electrode tip [14,15].

The carbon micro- and nanostructures could potentially be applied to the synthesis of electrodes for ion batteries [16], anode to electrochemical oxidation [17], light-emitting devices, organic transistors [18], catalyst carriers [19] or for other self-assembled nanostructures.

Polymer thin-films synthesized in low-temperature plasma have been found to be a material with a low dielectric constant applied to microelectronics [20], and as thin films with applications in gas sensors for the detection of ethanol and ammonia vapors of concentrations ranging from 100 to 1000 ppm [21] or as a thin hydrophobic layer [22]. These experiments were carried out in a glow discharge in a low-pressure atmosphere.

One of the most popular liquid hydrocarbons used as a carbon feedstock is cyclohexane. The products of decomposition of cyclohexane, such as ethylene and methane, have been widely used for the synthesis of CNT or diamond-like structures.

Huang et al. [23] showed that the deposition rate of a plasma-polymerized thin layer increases with a decrease in the H/C ratio. Cyclohexane is a hydrocarbon with a relatively low H/C ratio, equal to two, and, for this reason, was used in the experiments presented in this paper as a carbon feedstock for the synthesis of carbon microstructures in microdischarges. Additionally, the high vapor pressure (13 kPa at 298 K) of cyclohexane, and its high auto-pyrolysis temperature (>1000 K), predestinate this hydrocarbon as a source of carbon, which can be obtained at relatively low temperatures and at atmospheric pressures.

The aim of this paper is to investigate the process of synthesizing various carbon structures from cyclohexane vapors in high-voltage, low-current electrical microdischarges at atmospheric pressure between needle and plate electrodes. This configuration is typical for corona discharges, but the distance between the electrodes was shorter in order to obtain at least one linear dimension of the active plasma area in the range of the order of a few millimeters. This configuration allowed the initiation of microplasma in a limited region between the electrodes.

2. Materials and Methods

A schematic of the experimental set-up is shown in Figure 1. The experiments were carried out in an atmosphere of argon and cyclohexane as the carbon feedstock, at normal pressure, in a reactor chamber of 0.1 dm^3, made of PMMA. The hydrocarbon vapors were fed to the reactor from a bubbling flask, with argon as the carrier gas. The concentration of hydrocarbon in argon was controlled via mixing the cyclohexane vapors with additional argon. The flow rates of argon flowing directly to the reactor and throughout the flask were measured using flow meters. The temperature of liquid cyclohexane in the flask was stabilized at 21 $\pm$ 1 °C and was slightly lower than the room temperature to avoid vapor condensation on the pipe's walls. The concentration of cyclohexane vapors was calculated from the flow rates of the vapor of cyclohexane and argon. The total flow rate of the mixture was kept constant at a level of 0.051 dm^3/min. The concentration of cyclohexane was changed in the range of 1 to 5%. Cyclohexane of purity of 99.5% was supplied by Chempur (Poland) and argon of purity 99.999% by Linde gas. Before starting the experiment and plasma ignition, the reactor was rinsed with the gas mixture for 10 min to remove the air from the volume. Then, the voltage was switched on to the assumed magnitude set before. The voltage was switched off after a required time period.

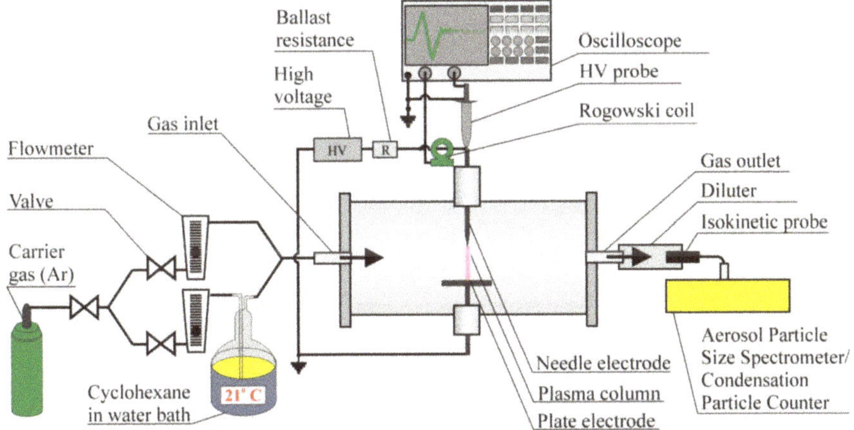

Figure 1. Experimental setup.

The electrical discharge was generated between a stainless-steel needle (composition in wt.%: Fe, 70; Cr, 18; Ni, 10; Si, 1) and a plate made of a nickel alloy (composition in wt.%: Ni, 75; Fe, 17; Mo, 5; Mn, 1.5). The diameter of the needle was 1 mm, the tip radius of the needle was about 30 µm. The dimensions of the plate were 35 × 20 mm. The distance between the electrodes was changed in the range of 1 mm to 15 mm.

The electrodes were supplied from high voltage power supply SPELLMAN HV SL 600 W/40 kV/PN (Spellman, Pulborough, UK) of a positive polarity. The discharge current was stabilized by a series resistance with a value of 5 MΩ. The discharge current was changed from 0.4 mA up to 3 mA. The voltage between the electrodes was measured with high voltage probe TEKTRONIX P6015A (Tektronix, Beaverton, OR, USA). The current pulses were measured by means of Rogowski type current monitor PEARSON 6600 (Pearson Electronics, Palo Alto, CA, USA) and Differential Preamplifier Tektronix ADA400A as amplifier of current signals. Both of the signals gained by these probes were recorded using digital storage oscilloscope TEKTRONIX TDS 3032.

The as-made carbon deposit grown on the needle tip or plate was carefully removed from the reactor, placed at the microscopic stage covered with carbon tape, and examined under a scanning electron microscope (Zeiss EVO 40, Carl Zeiss, Germany) equipped with an energy dispersive spectroscope (EDS, Quantax 200, Bruker, UK). Renishaw inVia Raman Microscope (Renishaw, UK) with a $100\times$ objective lens and a laser with a power of 1.5 mW at a wavelength of 514 nm was used in order to analyze the Raman spectra of obtained carbon structures. The concentration of nanoparticles generated in microdischarge and leaving the reactor was measured using GRIMM Condensation Particle Counter 5416 (GRIMM Aerosol Technik Ainring GmbH & Co. KG, Ainring, Germany). The size distribution of aerosol particles was determined using Aerosol Particle Size Spectrometer LAP 322 (Topas GmbH, Dresden, Germany).

3. Results

3.1. Discharge Modes

Five main modes of discharge in a gas at atmospheric pressure can be identified from the following current–voltage characteristics of electric discharges: onset streamer, glow corona, pulsed spark discharge and atmospheric pressure glow discharge [24,25].

The relationship between the voltage between the electrodes and the discharge current flowing through the gas depends on the kind of gas, the gas pressure, the distance between the electrodes, the geometry of the electrodes, and the electrode material. The current–voltage characteristics of electrical discharge in a needle–plate electrode system for a mixture of argon with cyclohexane at atmospheric pressure for different inter-electrode distances for a positive polarity and a cyclohexane concentration of 5% are shown in Figure 2.

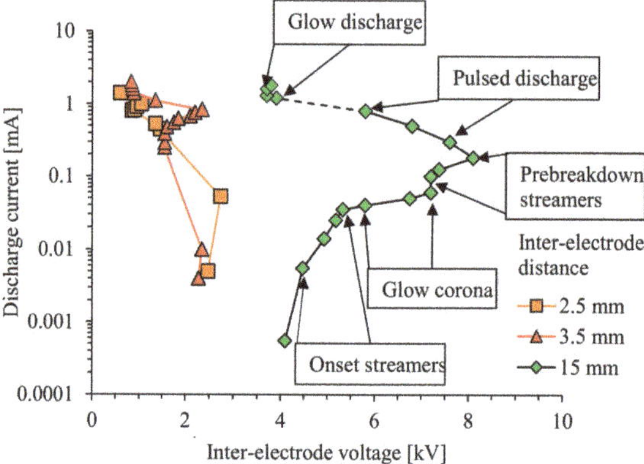

Figure 2. Current–voltage characteristics of electric discharge in Ar+C$_6$H$_{12}$ (95:5) mixture at atmospheric pressure for positive polarity of needle electrode, for three inter-electrode distances (2.5, 3.7 and 15 mm).

For a given magnitude of supply voltage, the time averaged discharge current was measured, and the mean value of the voltage drop between the electrodes was calculated using the following formula (see Figure 2):

$$Ud = Us - Id \cdot R, \tag{1}$$

where U_d is the voltage drop between the electrodes, Id is the discharge current, and R is the resistance of the external ballast resistor in the circuit.

Figure 3 shows photographs of various forms of electrical discharges in a mixture of argon and cyclohexane (95:5 by volume) for a positive polarity. All of the photographs were taken at the same optical magnification but with various exposures. The photographs show different stages of the development of the discharge with increasing voltage.

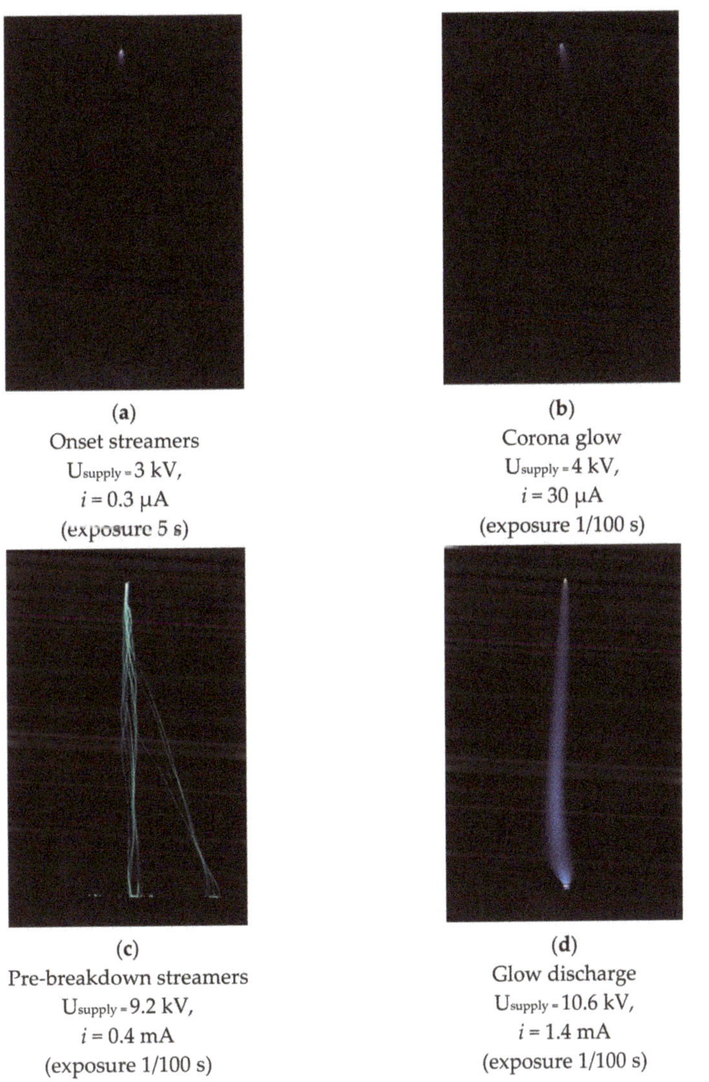

(a)
Onset streamers
$U_{supply} = 3$ kV,
$i = 0.3$ μA
(exposure 5 s)

(b)
Corona glow
$U_{supply} = 4$ kV,
$i = 30$ μA
(exposure 1/100 s)

(c)
Pre-breakdown streamers
$U_{supply} = 9.2$ kV,
$i = 0.4$ mA
(exposure 1/100 s)

(d)
Glow discharge
$U_{supply} = 10.6$ kV,
$i = 1.4$ mA
(exposure 1/100 s)

Figure 3. Photographs of various forms of corona discharge (**a**), onset streamers (**b**) corona glow (Hermstein glow), (**c**) pre-breakdown streamers and (**d**) atmospheric pressure glow discharge in a mixture of Ar+C_6H_{12} (95:5) for different supply voltages, for positive needle polarity (see [24]).

For voltages lower than the corona onset voltage, only dark discharge is generated. The discharge current is very low (<0.1 μA) because the electrons and ions flowing toward the electrodes, due to an applied electric field, are generated only by natural ionization. The ionization and excitation processes are so weak that the discharge is not visible to the "naked eye". This discharge was also not visible in the photos taken at a long exposure time of 15 s.

With the voltage increasing, the Townsend discharge appears in the needle region, and the current–voltage characteristics fulfil the quadratic Townsend law, for voltages $U_d > U_{d0}$, until the pre-breakdown streamers begin, as follows:

$$I_d = aU_d(U_d - U_{d0}), \tag{2}$$

where a is a constant depending on the configuration of electrodes, temperature, pressure and the kind of gas; I_d is the discharge current; U_d is the voltage between the electrodes, and U_{d0} is the corona onset voltage.

A corona discharge of positive polarity was visible as a faint glow at the anode tip, the area and intensity of which increased with an increasing supply voltage. Outside the luminous layer, the ionization processes are negligible—the number of electrons is insignificant, their kinetic energy is too low and only an ionic current flows toward the counter electrode (see [26,27]). This area is called the drift region. The kinetic energy of electrons is too low to sustain the excitation or ionization process in the drift region.

With the supply voltage increasing to about 7–8 kV, the pre-breakdown streamers appeared. The streamers are a spatially inhomogeneous phenomenon that occur over a period of several hundreds of nanoseconds. The value of the discharge current temporarily increased up to tens of microamperes from the lowest value for a given supply voltage.

A further increase in the supply voltage caused an increase in the discharge current to about 0.8 mA. Due to the voltage drop across the ballast resistance, the inter-electrode voltage decreased to about 5 kV. At this voltage, the electric field is still sufficiently high to produce a sufficient number of ions and photons bombarding the cathode to initiate the emission of secondary electrons from the negative electrode. At a potential leading to breakdown, the current may increase significantly, and a glow discharge or spark discharge could occur. The magnitude of the discharge current is limited by the ballast resistance in this case.

This form of discharge at atmospheric pressure was characterized by a filamentary plasma column that bridged two electrodes.

For pulsed discharge and atmospheric pressure glow discharges, carbon deposits started to grow on both electrodes and the growth rate depended on the discharge current. For a discharge current higher than 1.4 mA, a carbon deposit grew on the needle (anodic carbon deposit) toward the plate electrode. In Figure 4 is shown the growing process of the carbon deposit at the anode tip and the evolution of the plasma column in an atmospheric pressure glow discharge (the bright zones in the photographs). The contours of the discharge needle and the plate electrode surface are drawn with white lines. In these photographs, the tip of the carbon fiber stretching out from the needle tip (not visible in the photograph in Figure 5a) and the soot layer, or other carbonaceous materials at the plate electrode, are indicated.

The properties of deposited carbon structures synthesized in plasma depend not only on the gas properties, the carbon feedstock and discharge power, but also on the electric field distribution in the plasma column and the voltage drop near the electrodes. The magnitude of the electric field in the plasma column was determined by the measurement of voltage drop between the electrodes, which depends on the distance between the electrodes. This method is commonly used to determine the average voltage drop near the electrodes and the electric field magnitude [28–31].

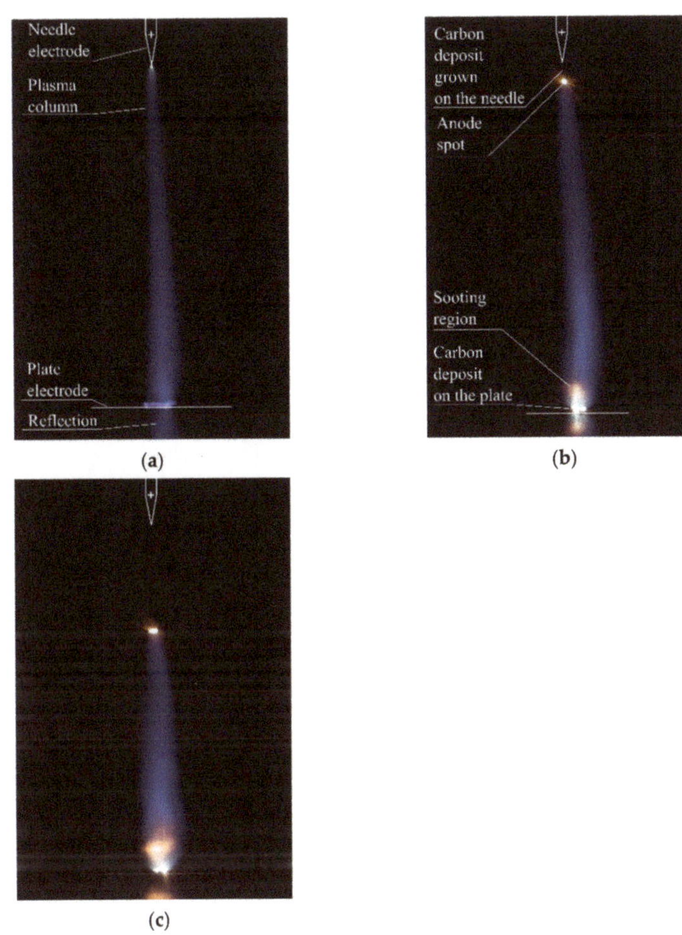

Figure 4. Photographs of plasma column of atmospheric pressure glow discharge during growth of carbon deposit in Ar+C_6H_{12} (95:5) for a discharge current $i = 1.78$ mA: (**a**) inception of glow discharge, (**b**) after 8 s and (**c**) after 26 s.

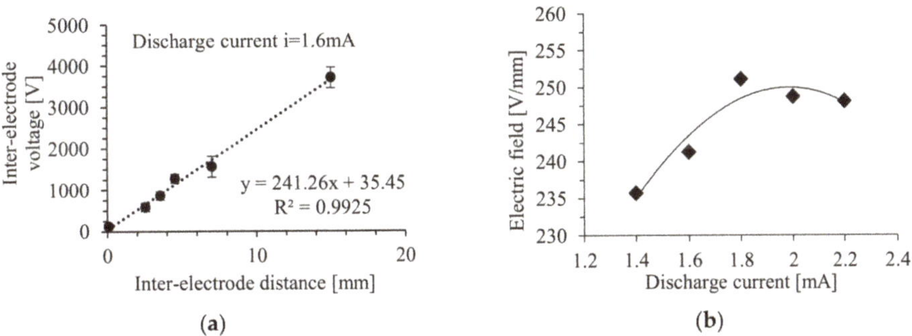

Figure 5. (**a**) Voltage drop between the electrodes for various inter-electrode distances, for discharge current $i = 1.6$ mA and (**b**) magnitude of electrical field for various discharge currents.

The distribution of the magnitude of the electric field in the plasma column was determined assuming that the voltage drop between the electrodes is the sum of the voltage drops across the anode and cathode and along the plasma column. By decreasing the distance between the electrodes, the voltage drop is also reduced due to the shorter length of the plasma column. The voltage drop near the electrodes is independent of the plasma column length, and the value of the voltage drop near the anode and cathode is constant regardless of the distance between them. Figure 5a shows the relationship between the voltage drop between the electrodes and inter-electrode distances, for a discharge current i = 1.6 mA. The sum of the voltage drops near electrodes can be determined by drawing a straight line, y = ax + b in this plot, the slope of which is the voltage drop across the plasma column, and the cutoff point, the sum of the anode and cathode drops. Figure 5b shows the magnitude of the electric field after a breakdown in the plasma column in a mixture of argon and cyclohexane versus the discharge current. The difference between the maximum electric field obtained for a discharge current of i = 1.8 mA and the lowest value for i = 1.4 mA, was not high (about 20 V), and the reduced electric field (E/N) was about 10 Td.

The average diameter of plasma column was estimated from the full width at half of the maximum brightness of the plasma column at photos taken for an exposure time of 1/2000 s. In a discharge between electrodes separated by an air gap of 3.7 and 15 mm, with a discharge current of 2.0 mA, the mean power density was estimated from the plasma column volume and the product of the discharge current and the inter-electrode voltage. The results are shown in Figure 6. The discharge power density slightly increases with the discharge current for large inter-electrode distances, but it decreases for small inter-electrode gaps. The highest power density was obtained for inter-electrode distances of 2.55 and 3.55 mm for all discharge currents (Figure 6).

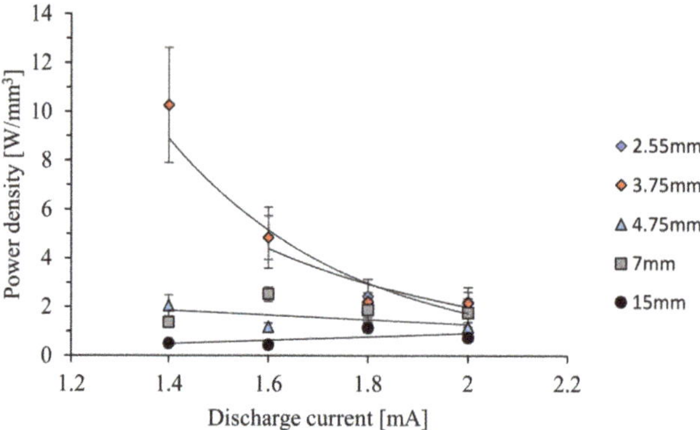

Figure 6. Power density of electrical discharge in a mixture of argon and cyclohexane (95:5 by vol.) obtained for various inter-electrode distances.

Increasing the power density in a plasma column increases the excitation energy of Ar, which could result in a more intense decomposition of cyclohexane. For temperatures above 600 °C the cyclohexane should start to be thermally decomposed, and ethane and butadiene are formed in the following reaction [32]:

$$c\text{-}C_6H_{12} \rightarrow C_2H_4 + C_4H_6 + H_2. \tag{3}$$

In the case of electrical discharges of energy above the metastable energy, the mainly $C_6H_{12}^+$ ions are generated after collision with electrons. If the energy will be higher at a level of Ar ionization energy, the main ionic product may be $C_4H_8^+$. The concentration of $C_5H_9^+$ ions is more than three times lower. In the latter case, CH_3 is a direct product of dissociation [33].

The power density vs. the discharge current in the case of a 15-millimeter inter-electrode distance for different cyclohexane concentration is shown in Figure 7. For a cyclohexane concentration in the range of 3 to 5%, the power density is nearly the same. For a 1% cyclohexane concentration, the power density was about two times higher for a discharge current in the range of 1.4–1.8 mA than for higher discharge currents.

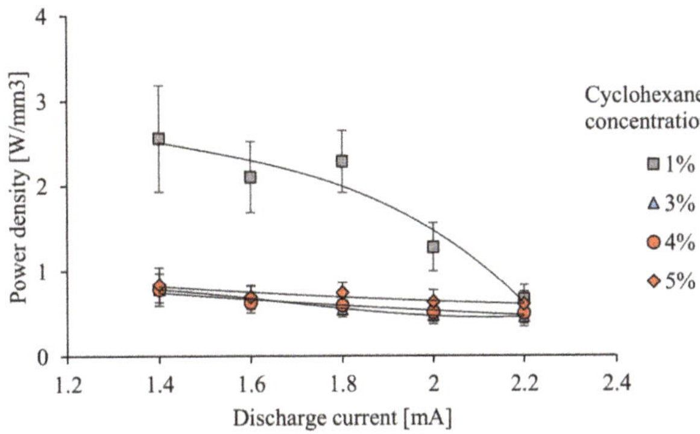

Figure 7. Power density of electrical discharge in a mixture of argon and cyclohexane vs. discharge current for different cyclohexane concentrations.

3.2. Discharge Diagnostics by OES

In this section, the results of optical emission spectroscopy (OES) investigations of glow discharge in a mixture of argon and cyclohexane are presented. The dependence of the concentration of activated species in the plasma and the composition of the working gasses are analyzed using OES in the wavelength range of 300–1000 nm. A typical OES result after breakdown is shown in Figure 8. The spectral lines and molecular bands were identified and assigned using online data available from the National Institute of Standards and Technology [34] and the book of Pearse and Gaydon [35], respectively. As can be seen, the infrared spectrum (696.5–978.5 nm) is dominated by the atomic spectrum of Ar optical emission lines, due to 1s-2p transitions (Table 1), whereas the visible spectrum (300–600 nm) is much more complex, but less intense, than the infrared part. A very weak emission at 516 nm was the C_2 Swan system. It should be noted that there is also a low-intensity emission band from N_2 (Second Positive System). The species identified using optical emission spectroscopy, which originated from cyclohexane decomposition, are listed in Table 2. The intensity of individual components varied depending on the discharge power. During the discharge, a continuous emission in the range of 350 to 450 nm was recorded. The continuous emission could be the effect of free-bound or free-free emission [36]. Free-bound continuous emission results from the capture of an electron by an ion, leading to recombination ($Ar^+ + e \rightarrow Ar^* + h\nu$). The second continuous emission is the effect of photon emission due to a decrease in the energy of an electron in the electric field of an ion or atomic nucleus ($Ar^+ + e \rightarrow Ar^+ + e + h\nu$, $Ar + e \rightarrow Ar + e + h\nu$, respectively). For a discharge current of 0.8 mA and higher, the continuous emission spectrum was observed in the range of 500–1000 nm, emitted by the deposit on the needle electrode.

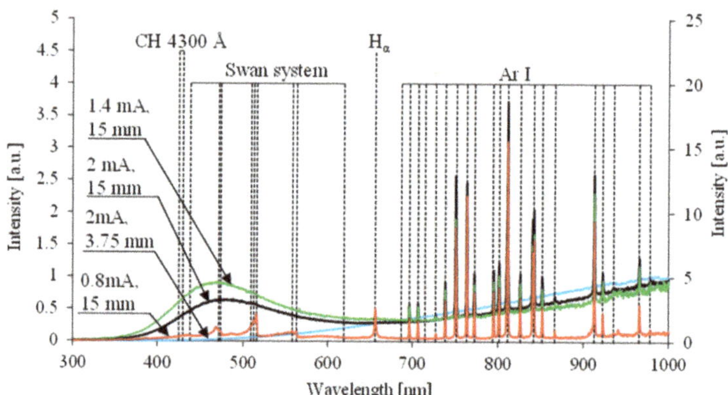

Figure 8. Optical emission spectra taken from the tip of needle electrode and from plasma column at 2 mm from the needle tip, for various discharge currents and inter-electrode distances.

Table 1. Optical emission lines of argon used for the determination of excitation temperatures during the discharge in Ar/C_6H_{12} mixture [34].

Transition (Upper Level Energy—Lower Level Energy [eV])	Position [nm]
$3s^23p^5(^2P^o_{1/2})4p \rightarrow 3s^23p^5(^2P^o_{3/2})4s$ (13.328–11.548)	696.5
$3s^23p^5(^2P^o_{1/2})4p \rightarrow 3s^23p^5(^2P^o_{3/2})4s$ (13.328–11.624)	727.3
$3s^23p^5(^2P^o_{1/2})4p \rightarrow 3s^23p^5(^2P^o_{3/2})4s$ (13.302–11.624)	738.4
$3s^23p^5(^2P^o_{3/2})4p \rightarrow 3s^23p^5(^2P^o_{3/2})4s$ (13.171–11.548)	763.5
$3s^23p^5(^2P^o_{1/2})4p \rightarrow 3s^23p^5(^2P^o_{1/2})4s$ (13.283–11.723)	794.8
$3s^23p^5(^2P^o_{1/2})4p \rightarrow 3s^23p^5(^2P^o_{1/2})4s$ (13.328–11.828)	826.5
$3s^23p^5(^2P^o_{1/2})4p \rightarrow 3s^23p^5(^2P^o_{3/2})4s$ (13.153–11.723)	866.8
$3s^23p^5(^2P^o_{3/2})4p \rightarrow 3s^23p^5(^2P^o_{3/2})4s$ (12.907–11.548)	912.3
$3s^23p^5(^2P^o_{3/2})4p \rightarrow 3s^23p^5(^2P^o_{1/2})4s$ (13.172–11.828)	922.5
$3s^23p^5(^2P^o_{3/2})4p \rightarrow 3s^23p^5(^2P^o_{3/2})4s$ (12.907–11.624)	965.8

Table 2. Main features of the emission spectra observed during cyclohexane decomposition in Ar/C_6H_{12} mixtures [35].

Species	System	Transition	Wavelength [nm]
C_2	Swan system	A3Π→X3Π, ground state	438–619
CH	4300 Å	A2Δ→X2Π, ground state	430
H	Balmer series	3→2	656.3

After the transition from pulsed to glow discharge, the intensity of all the optical emission lines decreases. An example of the evolution of the intensity of argon lines and C_2, CH and H bands with time, before and after the discharge onset, recorded with a time resolution of about 25 ms, is shown in Figure 9. After breakdown, a decrease in the intensity of the optical emission lines at 996.5 and 763.5 nm to about half of their maximum value (Figure 9a) was observed. The intensity of the C_2 Swan System for ν(0–0) after the onset of the discharge was much higher than the lines at 996.5 nm, resulting from excited argon, but after about 2 s, the intensity decreased to a very low level (Figure 9b). The emission intensities of the CH band and line from Hα were much lower than that of C_2.

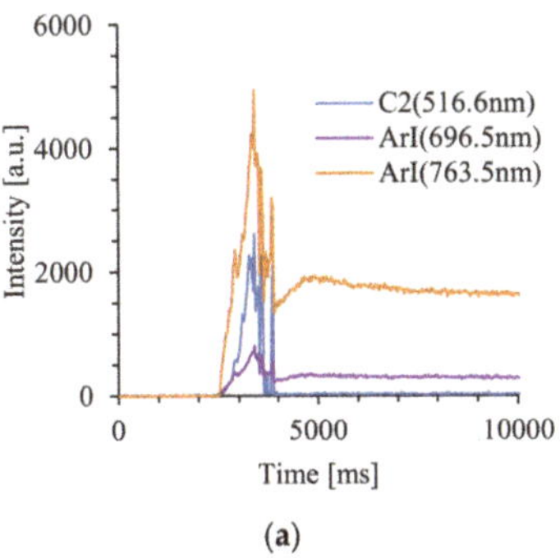

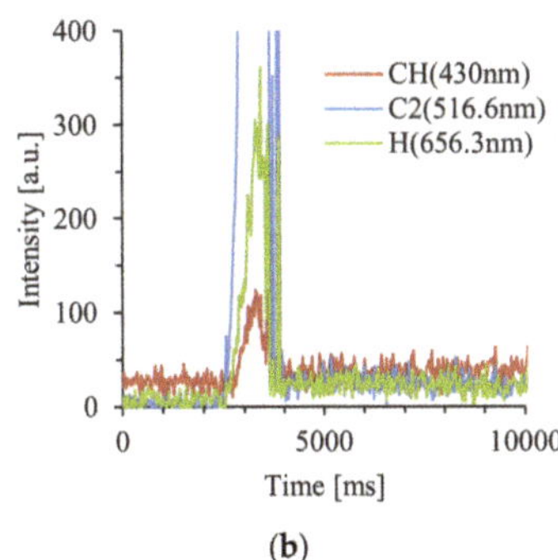

Figure 9. Time variation of the emission intensity of (**a**) C_2, ArI (at 696.5 nm) and Ar I (at 763.5 nm), and (**b**) CH, C_2 and Balmer series of hydrogen for a discharge current of $i = 1.8$ mA for a 15-millimeter inter-electrode distance.

The excitation energy of C_2 and CH is not very high (about 2.9 and 2.5 eV, respectively), but in order to obtain radicals from cyclohexane decomposition, much higher energy is required, and the chemical processes are much more complex. The most probable products, which could be obtained from cyclohexane decomposition, are ethane, methane and acetylene. For example, the emission of C_2 could be obtained by collision with electrons [37], as follows:

$$C_2H_2 + e \rightarrow C_2H + H + e, \tag{4}$$

$$C_2H + C_2H \rightarrow C_2 + C_2H_2, \tag{5}$$

or argon via an energy transfer from the metastable argon particles Ar*, as follows:

$$C_2H_2 + Ar^* \rightarrow C_2H + H + Ar, \tag{6}$$

$$C_2H + Ar^* \rightarrow C_2 + H + Ar. \tag{7}$$

For CH emission, the possible reactions could be as follows:

$$e + CH_2 \rightarrow CH + H + e, \tag{8}$$

$$e + CH_3 \rightarrow CH + 2H + e, \tag{9}$$

where CH_2 and CH_3 could be obtained by the decomposition of CH_4 [38] or C_2H_2 in electrical discharge [39].

Zhu et al. [40] measured the electron density in atmospheric pressure plasmas by comparing the ratio of the intensity of lines of argon 2p-1s transitions and interpreting them using a collisional–radiative model. The ratio of the intensity of lines of Ar (2p3 $\rightarrow$ 1s4) at 738.4 nm and Ar (2p6 $\rightarrow$ 1s5) at 763.5 nm is very sensitive to the variations of electron density in plasma in the range of 10^{14}–10^{16} cm^{-3}. By comparing these line ratios for different discharge currents and different inter-electrode distances, it can be concluded that the electron density in electrical discharge does not depend on the inter-electrode distance in the range of 2.55–15 mm. Additionally, there was not a significant difference between the discharge currents (Figure 10). The ratio of the intensity of the line was in the range of 0.25–0.3.

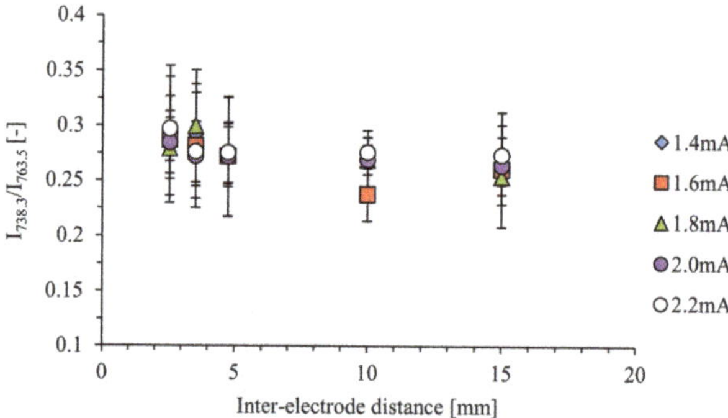

Figure 10. Intensity ratio of two lines of ArI for various discharge currents.

Atmospheric-pressure microplasmas are generally in near-partial-local thermodynamic equilibrium. The excitation temperature may be determined from Ar emission lines intensities obtained using the Boltzmann plot method from the following expression [41]:

$$ ln\left(\frac{I_{ij}\lambda_{ij}}{g_i A_{ij}}\right) = -\frac{E_i}{kT_{exc}} + C, \tag{10} $$

where I_{ij} is the relative intensity of the emission line between the energy levels i and j, λ_{ij} is its wavelength, g_i is the degeneracy or statistical weight of the emitting upper level i of the studied transition and A_{ij} is the probability of transition for a spontaneous radiative emission from the level, i, to the lower level, j. Finally, E_i is the excitation energy of the level i, k is the Boltzmann constant and C is a constant. An example of the linear Boltzmann plot for Ar I transition lines is shown in Figure 11.

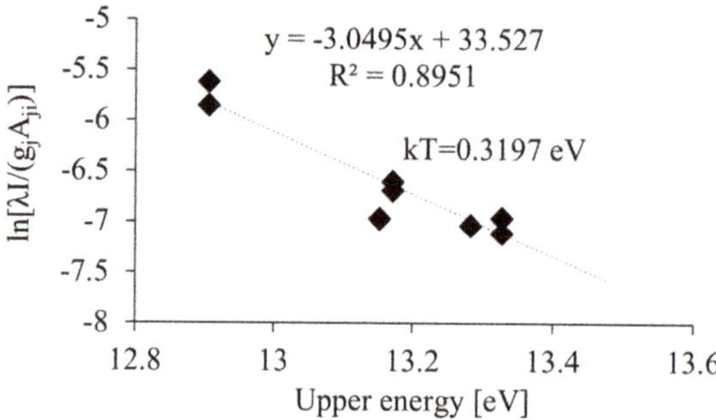

Figure 11. Linear Boltzmann plot for Ar I transition lines used to calculate excitation temperature (0.3197 eV).

Figure 12 shows the calculated excitation temperature as a function of discharge current for different inter-electrode distances. There is a difference between the excitation temperature for shorter inter-electrode distances and for longer inter-electrode distances. For a short inter-electrode distance (2.55 or 3.55 mm) the excitation temperature was about two times higher than for 15 mm.

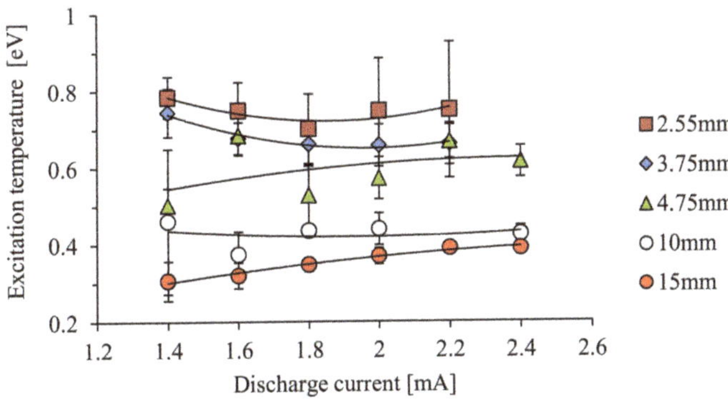

Figure 12. Excitation temperature of Ar I for various inter-electrode distances.

Assigning the continuous emission to the blackbody radiation of a hot anodic carbon deposit, the temperature of the deposit can be estimated through the fitting the continuous emission spectrum to the Planck blackbody radiation function. The anodic deposit temperature vs. discharge current for various inter-electrode distances is shown in Figure 13. The temperature of the deposit was estimated to be in the range of 1400 to 2200 K and is strongly dependent on the discharge current and inter-electrode distance. The temperature of the deposit also decreases with the concentration of cyclohexane (Figure 14). For the lowest concentration of cyclohexane (1%), the temperature of the anodic deposition was in the range of 1200–1600 K, which was much lower than for the deposit obtained for 3 and 5% cyclohexane. Additionally, the temperature of the deposit obtained for a negative polarity was lower, but it could be the effect of the deposit size (500 μm deposit diameter).

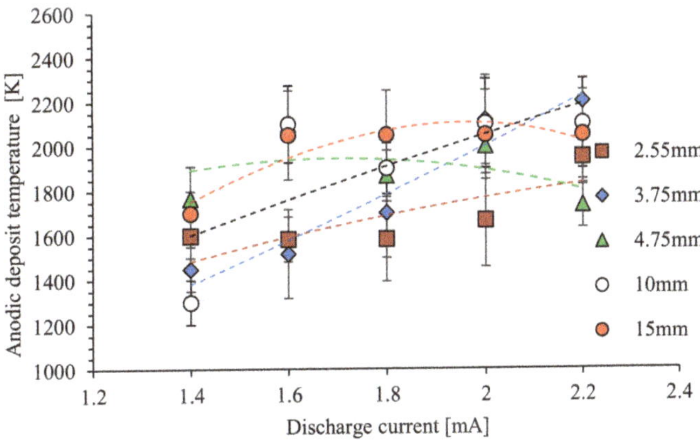

Figure 13. Anodic deposit temperature for various inter-electrode distances.

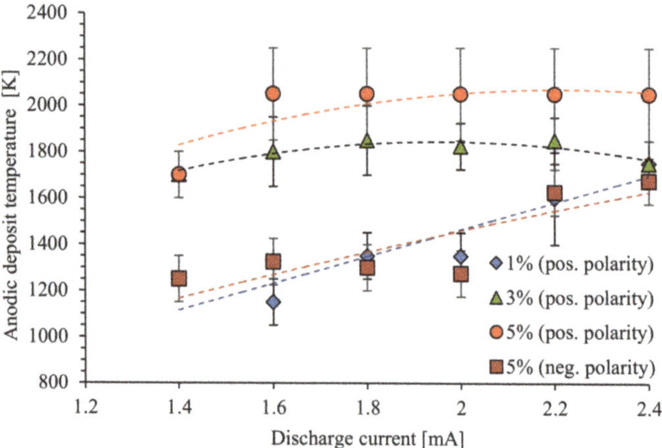

Figure 14. Temperature of needle deposit for positive polarity for 1 and 5% cyclohexane, and for negative polarity for 5% cyclohexane.

3.3. Carbon Deposit

Various forms of carbon deposit were obtained on the needle and plate electrodes as an effect of hydrocarbon decomposition in electric-discharge plasma. The morphology and growth rate of these deposits depended on the discharge current, the distance between the electrodes and the concentration of cyclohexane in the gas mixture.

In the case of onset streamers and glow corona discharges, the deposit obtained on the needle was in the form of irregular structures (Figure 15). The plate was coated by oiled substances during these irregular discharges, and some small carbon deposits in the form of a mound can also be obtained (photograph not shown). These kinds of structures were observed for an inter-electrode distance in the range of 2.55 to 25 mm.

Figure 15. Deposit formed during glow discharge for discharge current of 0.1 mA, in a mixture of Ar+C_6H_{12} (95:5 vol.).

Corona discharge is characteristic for a discharge where one of the electrodes is of high curvature. The magnitude of the electric field close to the electrode tip is large enough to lead to collisional ionization processes. Primary electrons released as a result of collisional ionization and secondary electrons released from the electrode by the bombardment of its surface by gaseous ions cause avalanche ionization in the presence of a strong electric field. These processes only occur in a thin layer near the electrode because there is a sufficiently strong electric field in this region (so-called ionization zone). The electric field in this area is of the order of magnitude of 106 V/m. With such a high electric field, the electrons present in this region are accelerated to a high kinetic energy sufficient for collisional ionization.

The main building blocks of carbon deposits obtained in the electric discharge microplasma in a mixture of argon and cyclohexane are radicals and neutral products of the decomposition of cyclohexane. These products are obtained due to the collision of electrons of sufficiently high energy with cyclohexane molecules; one example is the following:

$$c - C_6H_{12} + e \rightarrow R_m + R_n + e, \tag{11}$$

where R_m and R_n are radicals.

A similar situation occurs when an argon atom is in the first excited state (metastable state) after collision with an energetic electron and transfers its internal energy to a cyclohexane molecule after an impact, as follows:

$$c - C_6H_{12} + Ar_m \rightarrow Ar(^1S_0) + R_m + R_n. \tag{12}$$

As the energies of argon metastable states are relatively high, 11.55 eV and 11.72 eV, the cyclohexane and all products of cyclohexane decomposition can be ionized or dissociated to other products after collision. Radicals could interact with each other forming neutral molecules and/or other new radicals.

The formation of an "oil" form of deposit remaining on the plate electrode can proceed due to the interaction of radicals in the following reaction:

$$R_x + R_y \rightarrow P_{x+y}, \tag{13}$$

where P_{x+y} is a neutral product (hydrocarbon) of higher molecular weight than cyclohexane. The gas temperature in this zone is low, and the interaction with charged or high energetic particles is weak.

In the case of glow discharge, the interaction of atoms and molecules with electrons and charged particles occurs for all inter-electrode distances. Figure 16 shows typical SEM images of carbon deposits grown under different cyclohexane concentrations by a discharge current of 1.8 mA. The change of the cyclohexane concentration influenced the morphology of carbon fibers. For 1% cyclohexane, the surface of the deposited carbon fibers was grainy, with large spaces between the grains (Figure 16a). With an increasing concentration of cyclohexane, the edge of the grains appeared smoother and the space between the grains became much smaller than those deposited under a lower concentration of cyclohexane (Figure 16b). For a 3% cyclohexane concentration (Figure 16c), the structure of the carbon deposit has a tendency to become hybrid: one part of the deposit was grainy, and another part was smooth. For a 4 and 5% cyclohexane concentration, the surface of the carbon fiber was smooth (Figure 16d,e). For these concentrations of cyclohexane, soot particles were often stacked to the surface of the carbon deposit. The anodic carbon deposits have a diameter in the range of 30 to 100 μm, depending on the diameter of the plasma column.

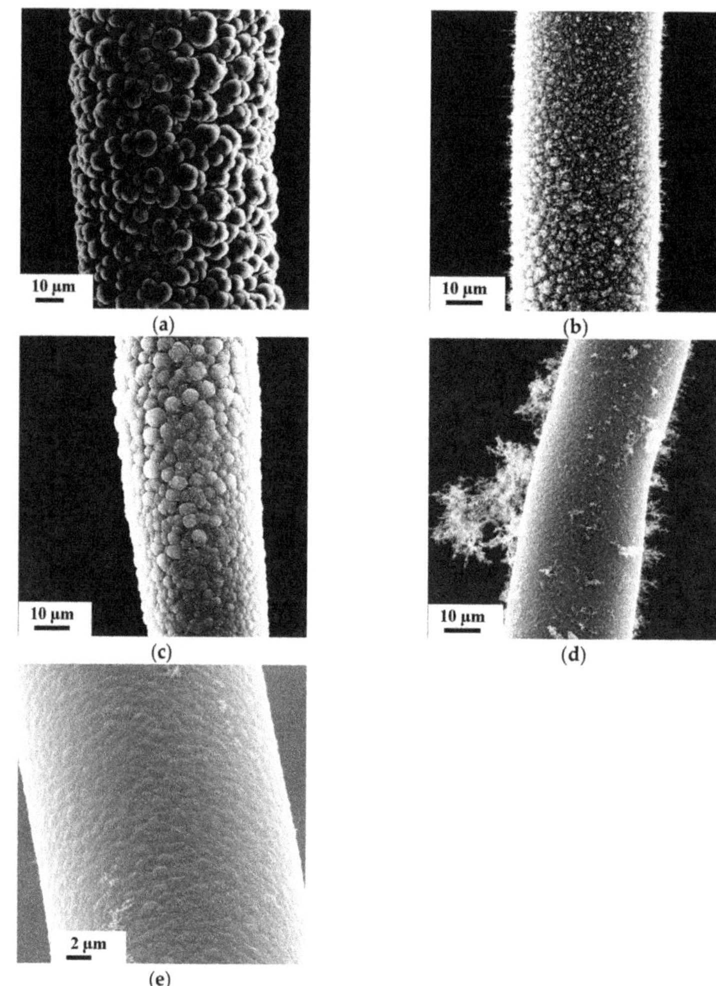

Figure 16. SEM pictures of carbon fibers obtained during corona discharge for a constant discharge current (1.8 mA) but for different C_6H_{12} concentrations: (**a**) 1%, (**b**) 2%, (**c**) 3%, (**d**) 4%, (**e**) 5% by positive polarity of the needle electrode. Electrode distance: 15 mm.

The growth rate of the carbon deposit was dependent on the discharge current (i.e., the power density delivered to the plasma column for cyclohexane decomposition) and the cyclohexane concentration. The volume growth rate of the carbon fibers deposited on the needle tip is shown in Figure 17. The growth rate for 1% cyclohexane was 3–4 times smaller than for 3 and 5% cyclohexane. For higher cyclohexane concentrations, the growth rate was similar, and was nearly a linear function of the discharge current.

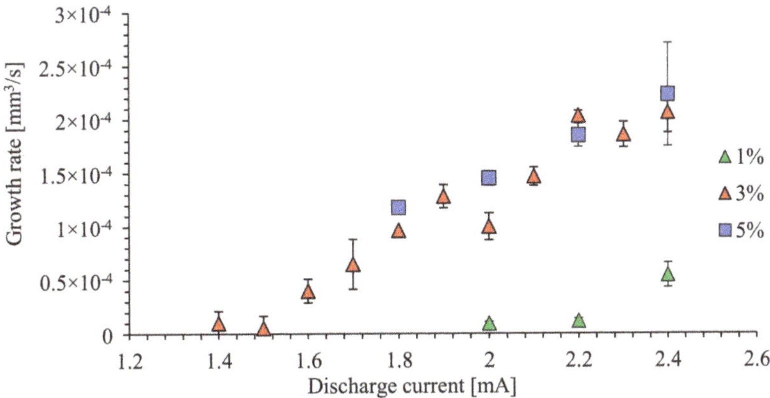

Figure 17. Growth rate of carbon deposit synthesized on the needle tip during electrical discharges versus discharge current, for three different cyclohexane concentrations: 1, 3 and 5%.

SEM images of the carbon deposit obtained on the cathode (needle electrode) during a negative polarity discharge in a mixture of argon $Ar+C_6H_{12}$ (95:5), by a discharge current of 1.8 mA are shown in Figure 18. For a negative polarity, the surface of the carbon deposit consisted of nodular dendrites, which have a diameter in the range of 5 to 10 μm, with large gaps between them. At a higher magnification, it is visible that each dendrite has cauliflower structures and consists of smaller nodules.

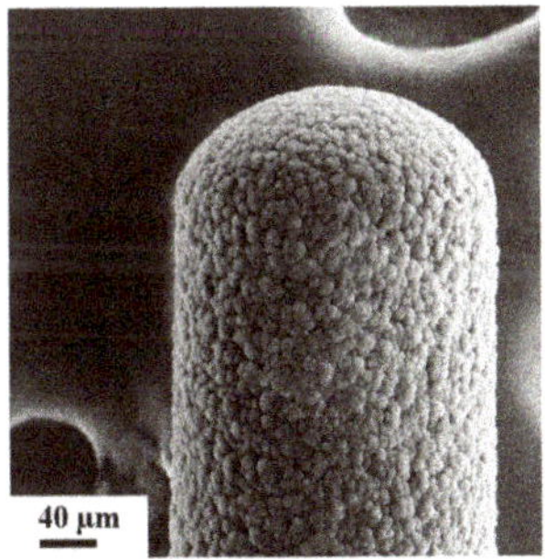
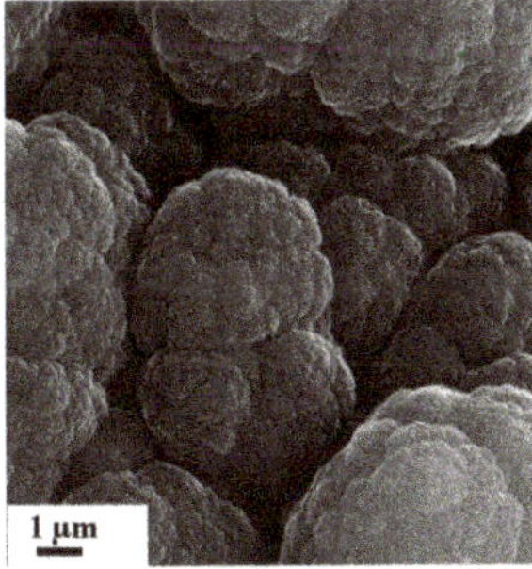

Figure 18. SEM pictures of cathodic carbon deposit obtained during discharge in a mixture of argon $Ar+C_6H_{12}$ (95:5) for negative polarity of the discharge electrode, with a discharge current of 1.8 mA.

Fragments of broken carbon deposits taken from the needle at a positive and negative polarity are shown in Figure 19. The deposits synthesized for low hydrocarbon concentrations and for a positive polarity of the discharge electrode are similar to those obtained for a negative polarity for 5% cyclohexane. In both cases, the carbon deposits have a columnar morphology with nodular dendrites, and the growth process is determined by the carbon particles' diffusion toward the electrodes (Peclet number >>1). Massuer et al. [42] suggested that the columnar morphology with voids between the column results from the process of columnar growth that occurs when the temperature of the deposit's surface is low and the energy of the ions bombarding the anodic deposit is low. The temperature determined from the thermal spectrum was about 1300 K, for the positive and negative polarities of the discharge electrode. The morphology obtained for a negative polarity, by 5% cyclohexane, could result from the diffusion-limited aggregation by low temperatures, due to the reduced surface mobility of adatoms and the high sticking probability of radicals and ions [43,44]. When a needle is used as a cathode, the main builders of the deposit could be positive ions, the main products of cyclohexane dissociation [45]. Additionally, the cathode sheath could influence morphology (the voltage drop of a cathode should be higher than for an anode).

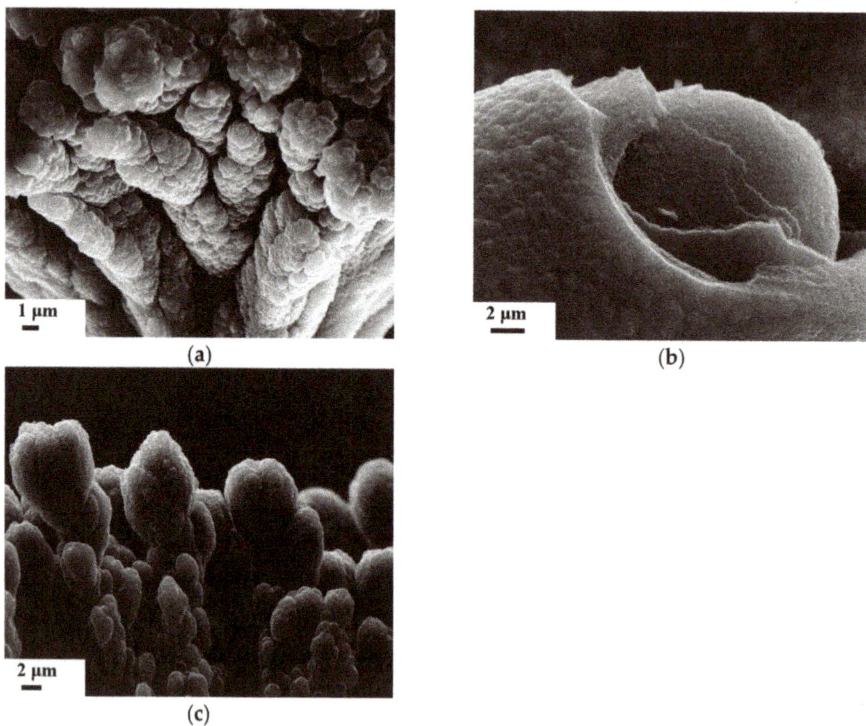

Figure 19. Fragments of carbon deposit obtained at the anode for cyclohexane concentrations of (**a**) 1%, (**b**) 5% for positive polarity and (**c**) 5% for negative polarity, with a discharge current of 1.8 mA.

The carbon deposit obtained for 5% cyclohexane is synthesized layer-by-layer with smooth morphology. This structure could be named a carbon fiber according to IUPAC nomenclature [46].

The SEM micrographs of the carbon deposits obtained for different discharge currents with an inter-electrode distance of 3.75 mm and a 5% cyclohexane concentration are shown in Figure 20. For a discharge current of 1.4 mA (Figure 20a), the carbon nanowalls were indistinct and barely visible. When the discharge current was increased to 1.6 mA (Figure 20b), the carbon nanowalls were large and densely packed. For a current of 1.8 mA (Figure 20c), carbon nanowalls are not produced and only the nanowalls' nuclei can be noticed on the surface of the deposit. For a discharge current of 2 mA (Figure 20d), carbon nanowalls did not grow.

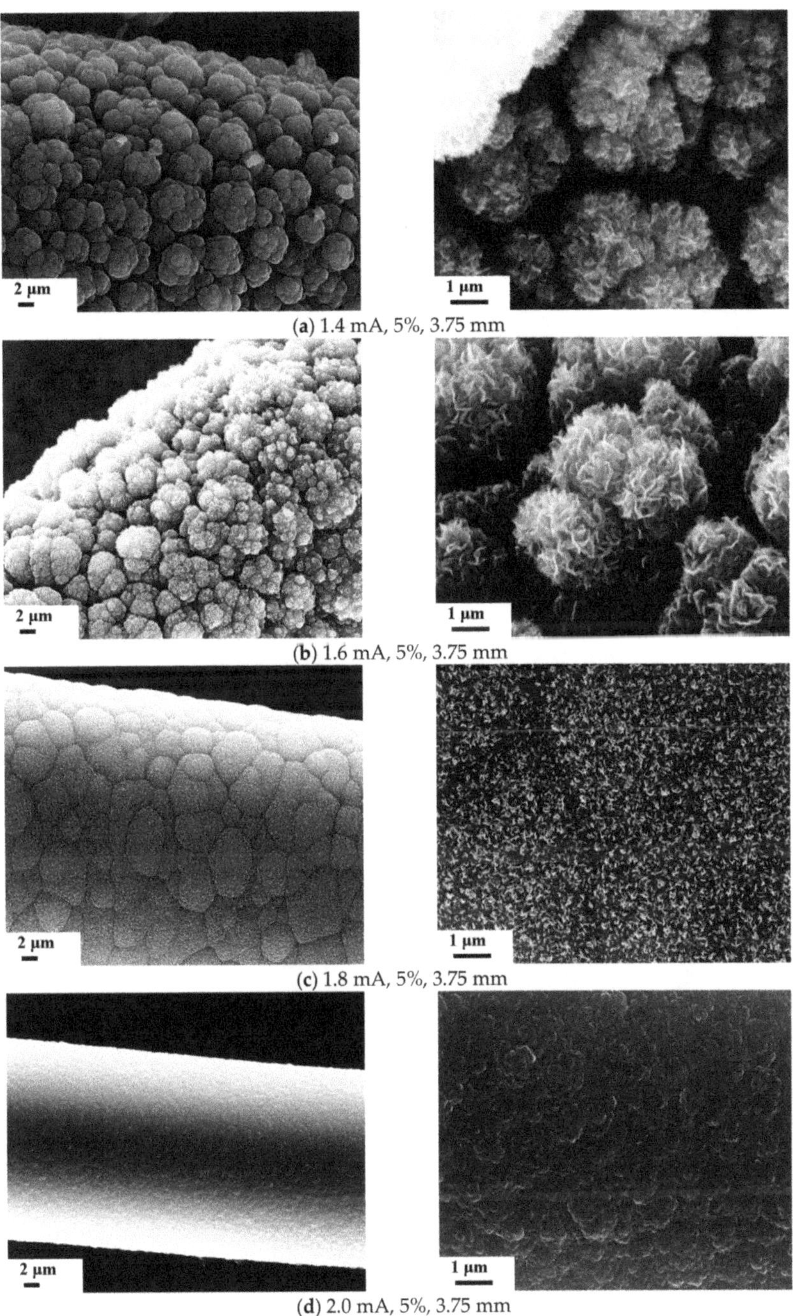

Figure 20. SEM pictures of carbon deposits obtained for different discharge currents: (**a**) 1.4 mA, (**b**) 1.6 mA, (**c**) 1.8 mA and (**d**) 2.0 mA with an inter-electrode distance of 3.75 mm and a 5% cyclo-hexane concentration. Left column: general view of carbon fiber, right column: close-up view of fiber's surface.

A comparison of the growth rate of the carbon deposit synthesized on the needle for different inter-electrode distances and various discharge currents is shown in Figure 21. The slowest growth rate was observed for the lowest inter-electrode distance. Generally, for an inter-electrode distance lower than 7 mm, the growth rate decreased, which could be an effect of the different radicals obtained during the decomposition of cyclohexane. The sticking probability of radicals such as CH_x, C_2H_x, CH and C differ in value [47], which could be dependent on the temperature of the surface/substrate [48].

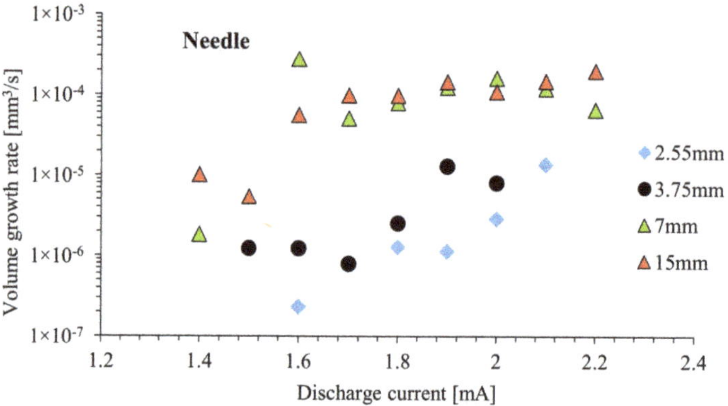

Figure 21. Growth rate of carbon deposit synthesized on the tip of needle electrode during electrical discharges of positive polarity versus discharge current, for a 5% cyclohexane concentration.

Figure 22 shows the current density flowing through the anode deposit determined from the discharge current divided by the average cross surface area of the carbon deposit at its tip. Increasing the current density results in the increasing temperature of the deposit surface due to Joule heating. The carbon deposits without carbon nanowalls had diameters in the range of 15–30 µm, whereas the diameters of the deposits covered with carbon nanowalls were in the range of 32–50 µm. The carbon deposits without carbon nanowalls and with carbon nanowalls were obtained for a current density higher than 395 A/cm² and lower than 320 A/cm², respectively.

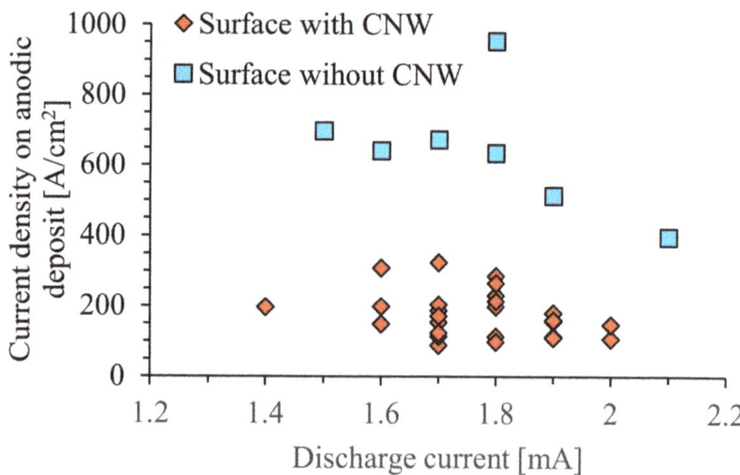

Figure 22. Current density on the anodic deposit for deposit with and without carbon nanowalls obtained for various discharge currents.

After deconvolution, the first-order Raman spectra of amorphous carbon material consist of two main peaks between 1200 and 1700 cm^{-1}. Similar results were obtained for the synthesized carbon structures and are shown in Figure 23a. The peak located near 1600 cm^{-1} is called graphitic (G band) and is associated with an in-plane stretching of sp2-bonded carbon atoms. The peak around 1350 cm^{-1} is called a disorder-induced or D band. For single-crystal perfect graphite D band is not observed [49].

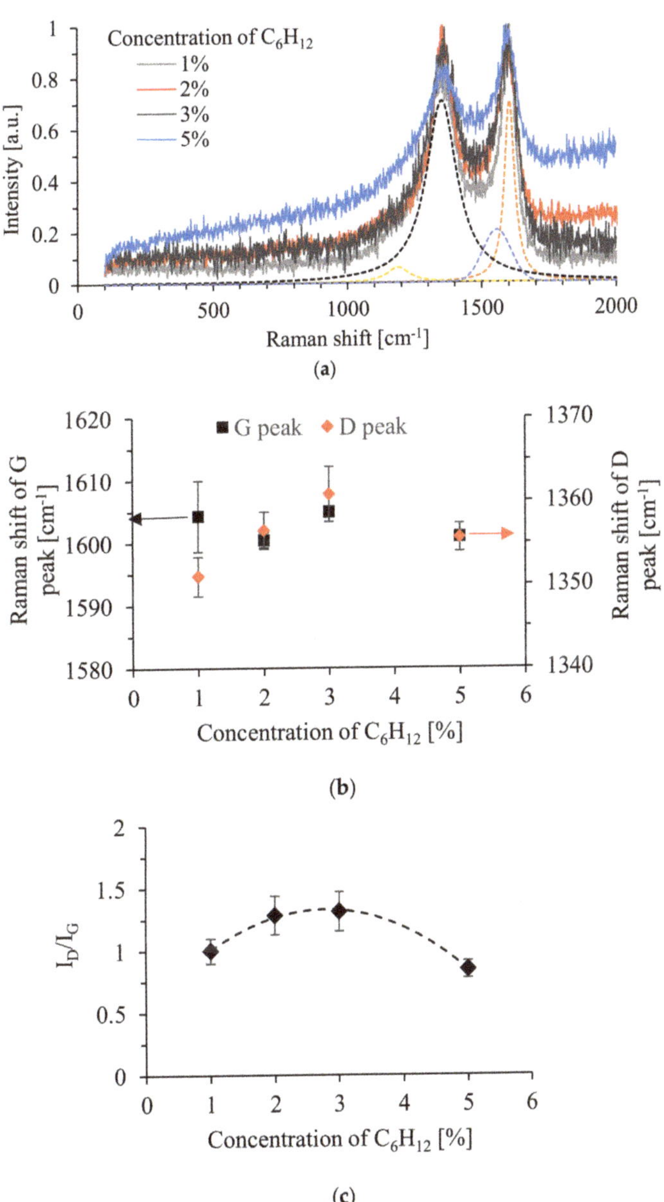

Figure 23. Raman spectra analysis of obtained carbon deposits: (**a**) Raman spectrum, (**b**) Raman shift of G and D bands, (**c**) the ratio intensity of D band and G band vs. cyclohexane concentration, with a discharge current of 1.8 mA.

Additionally, the obtained carbon fibers had both peaks, G and D band. For 1% cyclohexane, the G band was at about 1605 cm^{-1} and did not change significantly with an increasing cyclohexane concentration from 1 to 5 wt.% (Figure 23b). The position of the D band changed in the range of 1350 to 1360 cm^{-1}, with the maximum shift at 1360 cm^{-1} for 3% cyclohexane. In addition to the D and G bands, small peaks at 1200, 1510 and 1550 cm^{-1} were also obtained. The peak at 1200 cm^{-1} is attributed to the sp2-sp3 bonds or C-C and C=C stretching of vibrations of polyene-like structure [50] or vibrations of disordered graphite lattices (A1g symmetry) [51]. Both of the peaks, at 1200 and 1550 cm^{-1}, were observed in the deposits obtained for both polarities, positive and negative, and can be attributed to amorphous sp3-bonded carbon and sp2 carbon-based structures, respectively [52]. The peak at 1510 cm^{-1} is associated with the amorphous carbon content of soot [52] or C=C double-bond stretching vibrations [53]. It was also observed that the width of the G peak increased with an increasing cyclohexane concentration in the mixture, which could suggest an increase in the density of this deposit [54].

The degree of disorder of the carbon structures could be obtained from the relative intensities of the D- and G-peaks. Wopenka et al. [55] related carbon structures to the ratio of ID to IG, as follows: well-ordered graphite: ID/IG < 0.5, disordered graphite: 0.5 <I_D/I_G<1.1, glassy carbon I_D/I_G > 1.1. This suggest that carbon deposits obtained for 2 and 3% cyclohexane have a glassy carbon structure (Figure 23c). For a 1 and 5% cyclohexane concentration, the structures grown on the needle are disordered graphite. For a negative polarity, for a cyclohexane concentration of 5% and a discharge current $i = 1.8$ mA, the I_D/I_G ratio was about 1.7 (not shown).

An observed increase in I_D/I_G could result from the enhancement of the clustering of an aromatic ring in the deposited structures [56]. Additionally, an increasing luminescence background with an increasing cyclohexane concentration could result from a high content of hydrogen [57–59]. The average hydrogen content, analyzed using a total combustion method, of the carbon fibers obtained for $i = 1.8$ mA and 5% cyclohexane for a positive polarity was about 1.46 wt.%; therefore, it could be supposed that for a carbon anodic deposit obtained for a lower cyclohexane concentration, the hydrogen content should be lower than for 5% cyclohexane.

3.4. Carbon Aerosol Produced in the Discharge

Besides the carbon deposit produced at the surface of electrodes during electrical discharge in a mixture of cyclohexane and argon, aerosol particles, which leave the reactor with the flowing gas, were also generated. The variations in the concentration of particles produced during the discharge for two inter-electrode distances are shown in Figure 24. The particle concentration for discharge currents of $i = 1.4$ and $i = 2.4$ mA was similar and was higher than that for the discharge current of $i = 1.8$ mA. The total particle number concentration for the inter-electrode distance of 3.75 mm was about one order of magnitude higher than for 15 mm.

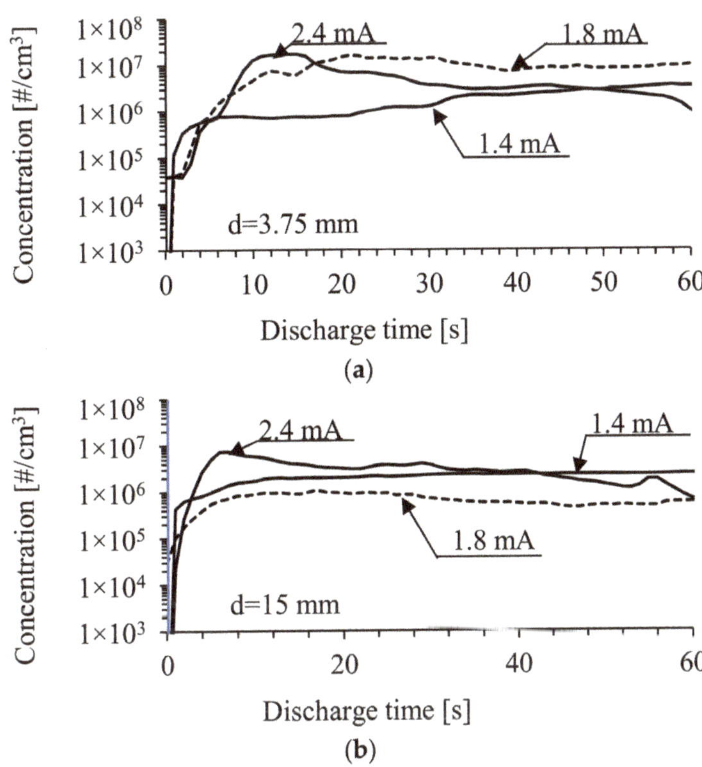

Figure 24. Concentration of particles produced during discharge in a mixture of argon and cyclohexane for two inter-electrode distances of (**a**) 3.75 mm and (**b**) 15 mm for different discharge currents.

The particle size distribution measured during discharge in a mixture of argon and cyclohexane is shown in Figure 25. The particles generated for an inter-electrode distance of 3.75 mm are smaller than those generated for 15 mm. For a discharge current of $i = 2.4$ mA and an inter-electrode distance of 15 mm, the concentration of particles larger than 300 nm was 10 times higher than for same discharge current for lower inter-electrode distances, which could be the effect of a higher concentration of soot precursor radicals. Shigeta and Murphy [60] showed that nanoparticle growth takes place on the outer edge of the plasma column.

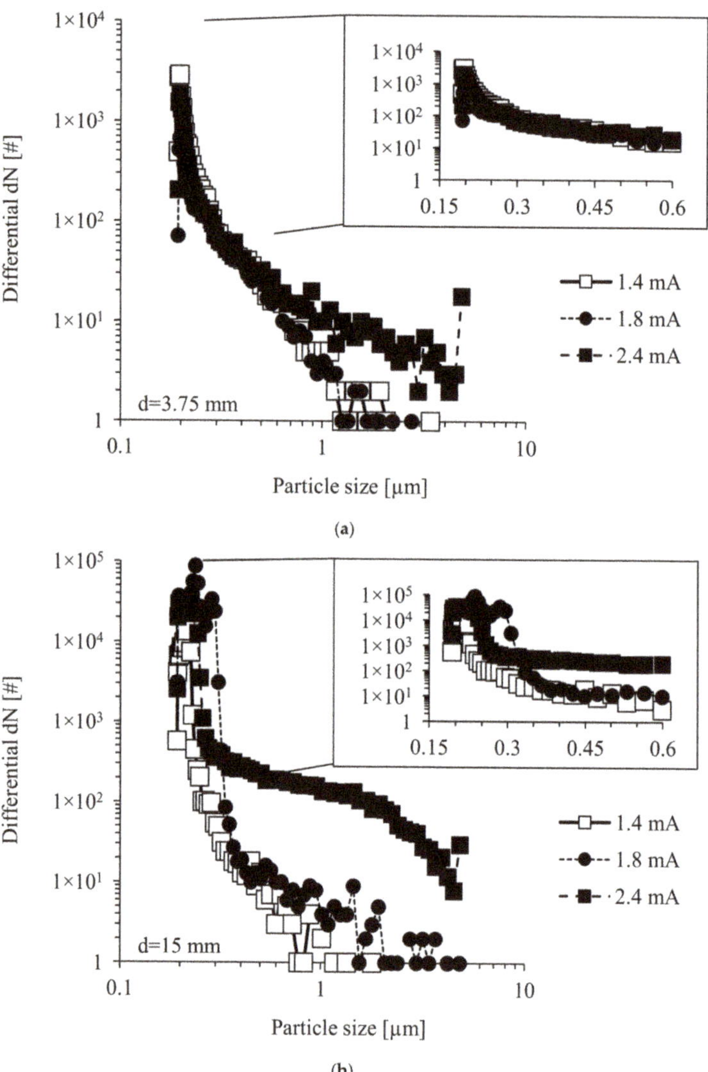

Figure 25. Particle size distribution produced during the discharge in a mixture of argon and cyclohexane, for inter-electrode distances of (**a**) 3.75 mm and (**b**) 15 mm for different discharge currents.

Soot can nucleate and grow through both ion-related and neutral radical mechanisms. Soot formation mostly occurs at the discharge boundary due to the relatively lower temperature than that in the plasma column. For example, for methane plasma, a sudden increase in the soot concentration occurs at temperatures above 1700 K [61]. In cyclohexane plasma, it could be the effect of increasing temperature and the acetylene produced in the discharge, which is known as a precursor of soot. The process of soot particle formation from acetylene, known as the "Winchester mechanism" [62], could be due to the following reactions with negative ions [63]:

$$C_2H_2 + e \rightarrow C_2H^- + H, \tag{14}$$

$$C_{2n}H_2 + C_2H \rightarrow C_{2n} + 2H^- + H_{2n}, \; n = 1,2,3 \ldots , \tag{15}$$

or

$$C_2H_2 + e \rightarrow H_2CC^-, \tag{16}$$

$$H_2CC^- + C_2H_2 \rightarrow C_{2n+2} + H_2^- + H_{2n}, \; n = 2,3,4 \ldots . \tag{17}$$

The same reactions could occur for positive ions.

For both inter-electrode distances, the mechanism should be the same, but the plasma column had a higher diameter when the interelectrode distance was 15 mm than when it was 3.75 mm.

4. Conclusions

The experimental results of the synthesis of carbon microstructures in low temperature plasma generated by microdischarges were presented. The experiments were carried out in a plasma reactor, between needle and plate electrodes, a configuration typical for corona discharge. As an effect of the decomposition of cyclohexane in plasma, thin, 50–100 μm in diameter, carbon microfibers were built at the tip of the needle electrode. Carbon nanowalls, carbon powder, soot particles and other structures were deposited onto these fibers. These various carbon structures, produced from cyclohexane vapors in high voltage, low-current electrical discharges at atmospheric pressure, were synthesized for different discharge currents, different electrode polarities, inter-electrode distances and cyclohexane concentrations. For 1% cyclohexane, the growth rate of the carbon deposit was very low compared to a higher concentration of this hydrocarbon, for the same discharge polarity and inter-electrode distance. The growth rate obtained for 3 and 5% cyclohexane, for 1.8 mA was similar, about 0.5 mm/s, but SEM investigations showed that the morphology of these deposits was different; they were semi grained and smooth surfaced, respectively. For an inter-electrode distance equal to and smaller than 4.55 mm, and a discharge current in the range of 1.4 to 1.8 mA, the growth rate was lower than for 15 mm, and the surface of the carbon deposits was covered by carbon nanowalls.

The process of deposition can, however, be controlled by the discharge current, the residence time in the reactor (flow rate), the gas composition and the geometry of electrode. The optimal conditions can only be determined experimentally.

Due to its porous structures, the produced carbon micro- and nanostructures have potential applications, for example, in the synthesis of electrodes for ion batteries, supercapacitors, fuel cells, gas sensor or catalyst carriers.

Author Contributions: Conceptualization, A.T.S. and A.J.; methodology, A.T.S.; validation, A.T.S. and A.J.; investigation, A.T.S.; writing—original draft preparation, A.T.S.; writing—review and editing, A.J.; supervision, A.J. Both authors have read and agreed to the published version of the manuscript.

Funding: This research was funded internally by the Institute of Fluid Flow Machinery, Polish Academy of Sciences within the project No. O2/T3/Z4.

Institutional Review Board Statement: Not applicable.

Informed Consent Statement: Not applicable.

Data Availability Statement: No supporting data.

Acknowledgments: The authors wish to thank Anna Białous for her technical help and for carrying out the Raman measurements.

Conflicts of Interest: The authors declare no conflict of interest.

References

1. Huczko, A. Synthesis of aligned carbon nanotubes. *Appl. Phys. A* **2002**, *74*, 617–638. [CrossRef]
2. Tachibana, K. Current status of microplasma research. *IEEJ Trans. Electr. Electron. Eng.* **2006**, *1*, 145–155. [CrossRef]
3. Martynov, Y.P.; Ivanov, V.A. Microdischarge of conducting particles and breakdown of a vacuum gap. *Radio Eng. Electron. Phys.* **1969**, *14*, 1732–1737.

4. Hara, M.; Akazaki, M. Analysis of microdischarge threshold conditions between a conducting sphere and plate. *J. Electrostat.* **1982**, *13*, 105–118. [CrossRef]

5. Gherardi, N.; Martin, S.; Massines, F. A new approach to SiO_2 deposit using a N_2 –SiH_4 –N_2O glow dielectric barrier-controlled discharge at atmospheric pressure. *J. Phys. D Appl. Phys.* **2000**, *33*, L104–L108. [CrossRef]

6. Shirai, H.; Kobayashi, T.; Hasegawa, Y. Synthesis of silicon nanocones using rf microplasma at atmospheric pressure. *Appl. Phys. Lett.* **2005**, *87*, 143112. [CrossRef]

7. Kozak, D.; Shibata, E.; Iizuka, A.; Nakamura, T. Growth of carbon dendrites on cathode above liquid ethanol using surface plasma. *Carbon* **2014**, *70*, 87–94. [CrossRef]

8. Kikuchi, T.; Hasegawa, Y.; Shirai, H. Rf microplasma jet at atmospheric pressure: Characterization and application to thin film processing. *J. Phys. D Appl. Phys.* **2004**, *37*, 1537–1543. [CrossRef]

9. Brock, J.R.; Lim, P. Formation of carbon fibers in corona discharges. *Appl. Phys. Lett.* **1991**, *58*, 1259–1261. [CrossRef]

10. Danilaev, M.P.; Bogoslov, E.A.; Morozov, O.G.; Nasybullin, A.; Pashin, D.M.; Pol′skii, Y.E. Obtaining Carbon Dendrites from the Products of Conversion of Polymer Materials. *J. Eng. Phys. Thermophys.* **2015**, *88*, 774–780. [CrossRef]

11. Li, M.-W.; Hu, Z.; Wang, X.-Z.; Wu, Q.; Chen, Y.; Tian, Y.-L. Low-temperature synthesis of carbon nanotubes using corona discharge plasma at atmospheric pressure. *Diam. Relat. Mater.* **2004**, *13*, 111–115. [CrossRef]

12. Sano, N.; Nobuzawa, M. Icicle-like carbon nanotubes forest at tungsten wire tip formed by high-voltage corona discharge. *Carbon* **2005**, *43*, 2224–2226. [CrossRef]

13. Mesko, M.; Vretenar, V.; Kotrusz, P.; Hulman, M.; Soltys, J.; Skaklaowa, V. Carbon nanowalls synthesis by means of atmospheric dcPECVD method. *Phys. Status Solidi B* **2012**, *249*, 2625–2628. [CrossRef]

14. Sobczyk, A.; Rajch, E.; Sozańska, M.; Jaworek, A. Formation of Carbon Fibres in High-Voltage Low-Current Electrical Discharges. *Solid State Phenom.* **2008**, *140*, 103–108. [CrossRef]

15. Sobczyk, A.T.; Jaworek, A. Carbon structures formation in low current high voltage electrical discharge in hydrocarbon vapours. *J. Phys. Conf. Ser.* **2011**, *301*, 012023. [CrossRef]

16. Yang, Q.; Wu, J.; Li, S.; Zhang, L.; Fu, J.; Huang, F.; Cheng, Q. Vertically-oriented graphene nanowalls: Growth and application in Li-ion batteries. *Diam. Relat. Mat.* **2019**, *91*, 54–63. [CrossRef]

17. Pierpaoli, M.; Jakobczyk, P.; Sawczak, M.; Łuczkiewicz, A.; Fudala-Książek, S.; Bogdanowicz, R. Carbon nanoarchitectures as high-performance electrodes for the electrochemical oxidation of landfill leachate. *J. Hazard. Mater* **2021**, *401*, 123407. [CrossRef]

18. Lee, J.-S.; Joh, H.-I.; Kim, T.-W.; Lee, S. Carbon nanosheets derived from soluble pitch molecules and their applications in organic transistors. *Org. Electron.* **2014**, *15*, 132–138. [CrossRef]

19. Lehmann, K.; Yurchenko, O.; Melke, J.; Fischer, A.; Urban, G. High electrocatalytic activity of metal-free and non-doped hierarchical carbon nanowalls towards oxygen reduction reaction. *Electrochim. Acta* **2018**, *269*, 657–667. [CrossRef]

20. Bae, I.-S.; Cho, S.-H.; Park, Z.T.; Kim, J.-G.; Boo, J.H. Organic polymer thin films deposited on silicon and copper by plasma-enhanced chemical vapor deposition method and characterization of their electrochemical and optical properties. *J. Vac. Sci. Technol. A* **2005**, *23*, 875. [CrossRef]

21. Antony, A.; Seo, H.J.; Boo, J. Plasma-polymerized cyclohexane coatings for ethanol and ammonia vapors sensing. *Asia-Pacif. J. Chem. Eng.* **2020**, *15*, 2452. [CrossRef]

22. Son, H.H.; Park, J.N.; Lee, W.G. Hydrophobic properties of films grown by torch-type atmospheric pressure plasma in Ar ambient containing C6 hydrocarbon precursor. *Korean J. Chem. Eng.* **2013**, *30*, 1480–1484. [CrossRef]

23. Huang, C.; Yu, Q.; Wu, S.-Y. Influence of the luminous gas phase on direct current plasma polymerized hydrocarbon film growth. *Vaccum* **2010**, *84*, 1402–1406. [CrossRef]

24. Sigmond, R.S. Simple approximate treatment of unipolar space-charge-dominated coronas: The Warburg law and the saturation current. *J. Appl. Phys.* **1982**, *53*, 891–898. [CrossRef]

25. Tachibana, K.; Koshiishi, T.; Furuki, T.; Ichiki, R.; Kanazawa, S.; Sato, T.; Mizeraczyk, J. A new measurement method of DC corona-discharge characteristics using repetitive ramp and triangular voltages. *J. Electrost.* **2020**, *108*, 103525. [CrossRef]

26. Goldman, M.; Sigmond, R.S. The corona discharge, its properties and specific uses. *Pure Appl. Chem.* **1985**, *57*, 1353–1362. [CrossRef]

27. Yanallah, K.; Pontiga, F. A semi-analytical stationary model of a point-to-plane corona discharge. *Plasma Sources Sci. Technol.* **2012**, *21*, 045007. [CrossRef]

28. Takaki, K.; Kitamura, D.; Fujiwara, T. Characteristics of a high-current transient glow discharge in dry air. *J. Phys. D Appl. Phys.* **2000**, *33*, 1369–1375. [CrossRef]

29. Titov, V.A.; Rybkin, V.V.; Maximov, A.I.; Choi, H.-S. Characteristics of Atmospheric Pressure Air Glow Discharge with Aqueous Electrolyte Cathode. *Plasma Chem. Plasma Process.* **2005**, *25*, 503–518. [CrossRef]

30. Pai, D.Z.; Lacoste, D.A.; Laux, C. Transitions between corona, glow, and spark regimes of nanosecond repetitively pulsed discharges in air at atmospheric pressure. *J. Appl. Phys.* **2010**, *107*, 093303. [CrossRef]

31. Hao, Y.; Zheng, B.; Liu, Y. Cathode fall measurement in a dielectric barrier discharge in helium. *Phys. Plasmas* **2013**, *20*, 113510. [CrossRef]

32. Fuge, R.; Liebscher, M.; Schröfl, C.; Damm, C.; Eckert, V.; Eibl, M.; Leonhardt, A.; Büchner, B.; Mechtcherine, V. Influence of different hydrocarbons on the height of MWCNT carpets: Role of catalyst and hybridization state of the carbon precursor. *Diam. Relat. Mat.* **2018**, *90*, 18–25. [CrossRef]

33. Jiao, C.Q.; Adams, S.F. Electron ionization of selected cyclohexanes. *J. Phys. B At. Mol. Opt. Phys.* **2011**, *44*, 175209. [CrossRef]

34. Kramida, A.; Ralchenko, Y.; Reader, J. *NIST ASD Team. NIST Atomic Spectra Database (Version 5.8)*; National Institute of Standards and Technology: Gaithersburg, MD, USA, 2019. Available online: https://physics.nist.gov/asd (accessed on 30 April 2021).

35. Pearse, R.W.B.; Gaydon, A.G. *The Identification of Molecular Spectra*, 3rd ed.; Chapman and Hall: London, UK, 1976.

36. Treshchalov, A.B.; Lissovski, A.A. VUV–VIS spectroscopic diagnostics of a pulsed high-pressure discharge in argon. *J. Phys. D Appl. Phys.* **2009**, *42*, 245203. [CrossRef]

37. Fantz, U.; Meir, S.; ASDEX Upgrade Team. Correlation of the intensity ratio of C_2/CH molecular bands with the flux ratio of C2Hy/CH4 particles. *J. Nucl. Mater.* **2005**, *337–339*, 1087–1091. [CrossRef]

38. Catherine, Y.; Pastol, A.; Athouel, L.; Fourrier, C. Diagnostics of CH, plasmas carbon deposition used for diamond-like carbon deposition. *IEEE Trans. Plasma Sci.* **1990**, *18*, 923–929. [CrossRef]

39. Gordillo-Vázquez, F.J.; Albella, J.M. Influence of the pressure and power on the non-equilibrium plasma chemistry of C2, C2H, C2H2, CH3 and CH4 affecting the synthesis of nanodiamond thin films from C2H2 (1%)/H_2/Ar-rich plasmas. *Plasma Sources Sci. Technol.* **2004**, *13*, 50–57. [CrossRef]

40. Zhu, X.M.; Pu, Y.K.; Balcon, N.; Boswell, R. Measurement of the electron density in atmospheric-pressure low-temperature argon discharges by line-ratio method of optical emission spectroscopy. *J. Phys. D Appl. Phys.* **2009**, *42*, 142003. [CrossRef]

41. Bashir, M.; Rees, J.M.; Bashir, S.; Zimmerman, W.B. Characterization of atmospheric pressure microplasma produced from argon and a mixture of argon–ethylenediamine. *Phys. Lett. A* **2014**, *378*, 2395–2405. [CrossRef]

42. Messier, R.; Giri, A.P.; Roy, R.A. Revised structure zone model for thin film physical structure. *J. Vac. Sci. Technol. A* **1984**, *2*, 500–503. [CrossRef]

43. Ren, Y.; Xu, S.; Rider, A.E.; Ostrikov, K. (Ken) Made-to-order nanocarbons through deterministic plasma nanotechnology. *Nanoscale* **2010**, *3*, 731–740. [CrossRef] [PubMed]

44. Michelmore, A.; Whittle, J.D.; Bradley, J.; Short, R.D. Where physics meets chemistry: Thin film deposition from reactive plasmas. *Front. Chem. Sci. Eng.* **2016**, *10*, 441–458. [CrossRef]

45. Lia, S.G.; Levin, R.D.; Kafafi, S.A. *Gas Phase Ion Energetics Data. NIST Chemistry WebBook, NIST Standard Reference Database Number 69*; National Institute of Standards and Technology: Gaithersburg, MD, USA, 2018. Available online: https://webbook.nist.gov (accessed on 30 April 2021).

46. McNaught, A.D.; Wilkinson, A.; IUPAC. *Compendium of Chemical Terminology*, 2nd ed.; Blackwell Scientific Publications: Oxford, UK, 1997.

47. Castro, M.; Cuerno, R.; Nicoli, M.; Vázquez, L.; Buijnsters, J. Universality of cauliflower-like fronts: From nanoscale thin films to macroscopic plants. *New J. Phys.* **2012**, *14*, 103039. [CrossRef]

48. Meier, M.; Von Keudell, A.; Jacob, W. Consequences of the temperature and flux dependent sticking coefficient of methyl radicals for nuclear fusion experiments. *Nucl. Fusion* **2002**, *43*, 25–29. [CrossRef]

49. Reich, S.; Thomsen, C. Raman spectroscopy of graphite. *Philos. Trans. R. Soc.* **2004**, *362*, 2271–2288. [CrossRef]

50. Ramya, K.; Jerin, J.; Manoy, B. Raman Spectroscopy Investigation of Camphor Soot: Spectral Analysis and Structural Information. *Int. J. Electrochem. Sci.* **2013**, *8*, 9421–9428.

51. Ivleva, N.P.; McKeon, U.; Niessner, R.; Pöschl, U. Raman Microspectroscopic Analysis of Size-Resolved Atmospheric Aerosol Particle Samples Collected with an ELPI: Soot, Humic-Like Substances, and Inorganic Compounds. *Aerosol Sci. Technol.* **2007**, *41*, 655–671. [CrossRef]

52. Wei, Q.; Ashfold, M.N.; Mankelevich, Y.; Yu, Z.; Liu, P.; Ma, L. Diamond growth on WC-Co substrates by hot filament chemical vapor deposition: Effect of filament–substrate separation. *Diam. Relat. Mater.* **2011**, *20*, 641–650. [CrossRef]

53. Iwaki, M.; Watanabe, H.; Matsunaga, M.; Terashima, K. Synthesis of diamond-like carbon structure by Naion implantation in graphite and polyacetylene. *Surf. Coat. Technol.* **1998**, *103–104*, 370–374. [CrossRef]

54. Chu, P.K.; Li, L. Characterization of amorphous and nanocrystalline carbon films. *Mater. Chem. Phys.* **2006**, *96*, 253–277. [CrossRef]

55. Wopenka, B.; Xu, Y.; Zinner, E.; Amari, S. Murchison presolar carbon grains of different density fractions: A Raman spectroscopic perspective. *Geochim. Cosmochim. Acta* **2013**, *106*, 463–489. [CrossRef]

56. Nakao, S.; Choi, J.; Kim, J.; Miyagawa, S.; Miyagawa, Y.; Ikeyama, M. Effects of positively and negatively pulsed voltages on the microstructure of DLC films prepared by bipolar-type plasma based ion implantation. *Diam. Relat. Mater.* **2006**, *15*, 884–887. [CrossRef]

57. Yoshikawa, M.; Katagiri, G.; Ishida, H.; Ishitani, A.; Akamatsu, T. Raman spectra of diamondlike amorphous carbon films. *J. Appl. Phys.* **1988**, *64*, 6464–6468. [CrossRef]

58. McConnell, M.L.; Dowling, D.P.; Pope, C.; Donnelly, K.; Ryder, A.G.; O'Connor, G.M. High pressure diamond and dia-mond-like carbon deposition using a microwave CAP reactor. *Diam. Relat. Mat.* **2002**, *11*, 1036–1040. [CrossRef]

59. Ferrari, A.C.; Robertson, J.; Ferrari, A.C.; Robertson, J. Interpretation of Raman spectra of disordered and amorphous carbon. *Phys. Rev. B* **2000**, *61*, 14095–14107. [CrossRef]

60. Shigeta, M.; Murphy, A. Thermal plasmas for nanofabrication. *J. Phys. D Appl. Phys.* **2011**, *44*, 174025. [CrossRef]

61. Kouprine, A.; Gitzhofer, F.; Boulos, M.; Fridman, A. Polymer-Like C:H Thin Film Coating of Nanopowders in Capacitively Coupled RF Discharge. *Plasma Chem. Plasma Process.* **2004**, *24*, 189–215. [CrossRef]

62. De Bleecker, K.; Bogaerts, A.; Goedheer, W. Detailed modeling of hydrocarbon nanoparticle nucleation in acetylene discharges. *Phys. Rev. E* **2006**, *73*, 026405. [CrossRef]

63. Van De Wetering, F.M.J.H.; Beckers, J.; Kroesen, G.M.W. Anion dynamics in the first 10 milliseconds of an argon–acetylene radio-frequency plasma. *J. Phys. D Appl. Phys.* **2012**, *45*, 485205. [CrossRef]

 *applied sciences*

Article

Selective Plasma Etching of Polymer-Metal Mesh Foil in Large-Area Hydrogen Atmospheric Pressure Plasma

Richard Krumpolec *, Jana Jurmanová, Miroslav Zemánek, Jakub Kelar, Dušan Kováčik and Mirko Černák

Department of Physical Electronics, CEPLANT—R&D Center for Plasma and Nanotechnology Surface Modifications, Faculty of Science, Masaryk University, Kotlářská 267/2, 61137 Brno, Czech Republic; janar@physics.muni.cz (J.J.); mzemanek@mail.muni.cz (M.Z.); jakub.kelar@mail.muni.cz (J.K.); dusan.kovacik@mail.muni.cz (D.K.); cernak@physics.muni.cz (M.Č.)
* Correspondence: krumpolec@mail.muni.cz

Received: 29 September 2020; Accepted: 19 October 2020; Published: 21 October 2020

Featured Application: Surface selective etching of polymer substrates, metal-polymer composites, e.g., transparent conductive substrates by atmospheric plasma roll-to-roll processing.

Abstract: We present a novel method of surface processing of complex polymer-metal composite substrates. Atmospheric-pressure plasma etching in pure H_2, N_2, H_2/N_2 and air plasmas was used to fabricate flexible transparent composite poly(methyl methacrylate) (PMMA)-based polymer film/Ag-coated Cu metal wire mesh substrates with conductive connection sites by the selective removal of the thin (~10–100 nm) surface PMMA layer. To mimic large-area roll-to-roll processing, we used an advanced alumina-based concavely curved electrode generating a thin and high-power density cold plasma layer by the diffuse coplanar surface barrier discharge. A short 1 s exposure to pure hydrogen plasma, led to successful highly-selective etching of the surface PMMA film without any destruction of the Ag-coated Cu metal wires embedded in the PMMA polymer. On the other hand, the use of ambient air, pure nitrogen and H_2/N_2 plasmas resulted in undesired degradation both of the polymer and the metal wires surfaces. Since it was found that the etching efficiency strongly depends on the process parameters, such as treatment time and the distance from the electrode surface, we studied the effect and performance of these parameters.

Keywords: hydrogen plasma; atmospheric pressure plasma; selective etching; polymer-metal mesh composite foil; roll-to-roll processing

---

1. Introduction

Plasma technologies using a wide scale of plasma working gases has succeeded as a powerful tool for surface cleaning, functionalization, adhesion improvement, deposition, sputtering and etching. Hydrogen plasma is a very strong reducing agent already tested for the surface treatments of metals, silicon, carbon, and polymer materials. A comprehensive summary of plasma etching of polymers and polymeric materials including the hydrogen plasma etching can be found in the review of Puliyalil and Cvelbar [1]. In general, when polymer surfaces are exposed to the hydrogen plasma, the bombardment by plasma species (electrons, hydrogen ions and radicals) led to the degradation of the surface and etching reactions will occur.

The effect of hydrogen content in low-pressure N_2/H_2 plasma for surface modification of polyethylene (PE), poly(tetrafluoroethylene) (PTFE) and polyvinylidene fluoride (PVDF) studied Sara-Bournet et al. [2]. Hydrogen-containing plasma was found to causing dehydrohalogenation along the fluoropolymer backbone [3]. Subsequently, the co-process gas can covalently attach to these sites; thus, the hydrogen plasma with a co-process gas can particularly improve the adhesive bonds of fluoropolymers. Selective removal of polymeric PMMA residues on monolayer graphene by high-density, low pressure H_2 and H_2/N_2 plasmas without a damage to the graphene surface as examined by Cunge et al. [4]. Selective etching of graphene edges and graphene nanoribbons using hydrogen plasma reaction at 300 °C and at a pressure of 300 mTorr was studied by Xie et al. [5]. The advantage of selective plasma etching using hydrogen as etchant gas was used for selective preparation of metallic and semiconducting single-wall carbon nanotubes by Hou et al. [6] and Zhang et al. [7].

Inline plasma surface treatments showed a significant potential for applications in high volume and cost-effective roll-to-roll (R2R) fabrication of flexible electronics on a polymer web substrate including, for example, flexible solar cell and wearable electronics fabrication. Virtually all these applications need diffuse non-thermal plasma. We use the wording "diffuse plasma" in accordance with the term published for the first time by Šimor et al. in 2002 [8]. The "diffusivity" of the plasma is related to the visual uniformity of the generated plasma as observed by the human eye. So that we used the term diffuse plasma to refer to the plasma which, when visually observed, is widely spread, and not concentrated to individual plasma filaments. In our experience, the diffuse, visually uniform plasma enables uniform plasma surface modification working at the plasma exposure times longer than the time resolution of a human eye.

However, such plasma generated at low pressure in costly vacuum systems is at odds with high throughput processing required in industry [9]. However, the developments in the field of plasma physics during the last two decades have opened a way how to generate stable, diffuse and non-thermal plasma at atmospheric pressure [10]. The research of atmospheric plasma, including the atmospheric-pressure H_2 plasma, has been driven by the need of replacing of expensive vacuum systems by simpler, more cost-effective and high-throughput systems [9]. For example, [11] studied plasma etching of various polymers like PP, PET, PEN, PEEK, PMMA, PLA, LDPE and Nylon 6.6 using atmospheric pressure hydrogen plasma. Kwon et al. [12] used hydrogen plasma sintering process at temperature 150 °C for reduction and densification of copper complex ink on flexible PET substrate.

The so-called Diffuse Coplanar Surface Barrier Discharge (DCSBD) is capable to generate visually an almost uniform diffuse ambient air plasma with effective thickness of approximately 0.25 mm [13]. Large-area atmospheric-pressure hydrogen plasma generated by DCSBDs in a geometry mimicking R2R treatment was studied for fast plasma reduction of graphene oxide film on flexible PET substrate [14]. It was also reported that DCSBD hydrogen plasma can be used for chemical reduction of a native surface CuO/CuO_2 layers [15] as well as for etching of thin SiO_2/Si films [16] and reduction of diamond nanoparticles [17]. Reducing DCSBD plasma was already used for surface modification [18] and nanostructuring of PMMA, PET and PTFE polymeric substrates [19].

In this paper hydrogen atmospheric pressure plasma generated by a DCSBD with the electrode geometry applicable for the R2R processing was used for etching of thin surface PMMA polymer layer from the top of polymer-metal mesh substrate used as a flexible transparent conductive substrate. The etching was studied by scanning electron microscopy (SEM) and using a laser confocal microscope. A chemical reduction of plasma-modified surfaces was analyzed by energy dispersive x-ray analysis (EDX). The effect of process parameters on the properties of etched surface was investigated too. The selectivity of plasma etching was studied through the etching in mixtures of nitrogen and hydrogen. DCSBD can generate atmospheric plasma in all technically important gases, such as ambient air, nitrogen and hydrogen. Since the use of pure atmospheric pressure hydrogen can be of some safety concern, we also tested the etching capability and selectivity of some other technically and economically helpful plasma gases.

The novelty of this work is that the reducing, large-area hydrogen plasma generated at atmospheric pressure was applied to a polymer-metal mesh substrate in order to carry out selective plasma etching together with the optimization of etching parameters, in particular: plasma-substrate distance, etching time and working gas specification. The selectivity of plasma etching is of great importance especially in case of processing of complex polymer-metal composite materials.

2. Materials and Methods

Plasma etching was performed on a composite mesh foil (Sefar AG, Thal, Switzerland), consisting of a grid of polymer and metal wires embedded in a polymer PMMA foil. The thickness of the mesh foil was 100 µm. Metal wires, from copper, of thickness 40 µm were coated with a thin 1 µm thick layer of silver. The metal wires embedded in the foil were oriented longitudinally to the machine direction of the plasma treater.

Figure 1 shows a reactor designed for roll-to-roll treatment and based on diffuse coplanar surface barrier discharge with a concavely curved alumina ceramics used to etch polymer mesh samples by non-thermal, low temperature (<80 °C) atmospheric pressure plasma. A 0.3-mm thin plasma layer with surface power density 2.5 W cm^{-2} was generated in pure nitrogen and pure hydrogen (both 99.998%, Messer Technogas, Prague, Czech Republic) and its mixtures (5% and 50% of H_2 in N_2) at a flow rate of 3.5 L/min. Plasma treatment in ambient air was done in a closed reactor without the flow of working gas. The discharge was fed by a sinusoidal high-frequency high voltage (15 kHz) signal at input power 400 W. The area of a thin layer of DCSBD plasma was 195 mm × 80 mm. To mimic a roll-to-roll plasma etching conditions, the sample of 15 cm × 20 cm size was attached to the metal roller covered by a rubber with a diameter of 296 mm. Plasma treatment time in the range 0.25–3 s was changed by the setting of circumferential speed of the roller. The distance H between the sample surface and the concave DCSBD electrode surface was set to 0.35 mm. The samples were studied using SEM, EDX and confocal microscopy.

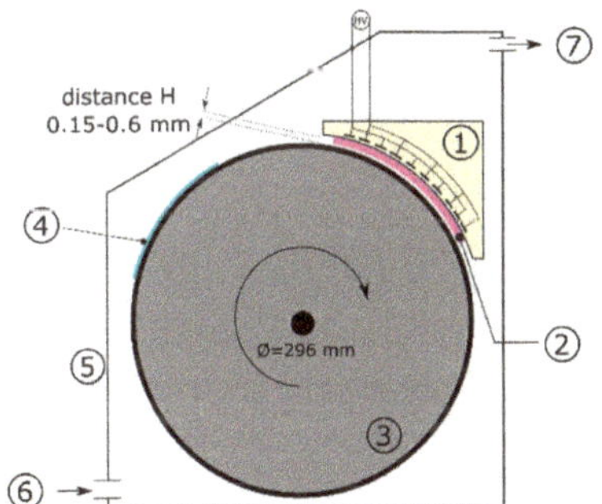

Figure 1. Reactor based on diffuse coplanar surface barrier discharge (DCSBD) with concavely curved alumina ceramics for roll-to-roll treatment of flexible substrates. 1—Concavely curved DCSBD unit; 2—DCSBD plasma; 3—roller; 4—sample; 5—reactor chamber; 6—gas inlet; 7—gas outlet.

Confocal Laser Microscope (LEXT OLS4000 3D Laser Measuring Microscope) was employed to observe the changes on plasma etched samples. Scanning Electron Microscopy with Energy dispersive X-ray Analysis using a MIRA3 device (TESCAN, Brno, Czech Republic) was used to reveal the changes in surface morphology, and chemical changes of plasma treated samples. SEM micrographs were taken using the accelerating voltage 10 kV; the edge of the samples was observed at accelerating voltage 30 kV. All samples were coated with a 10-nm layer of Au before the SEM analysis.

3. Results

3.1. Plasma Etching of Thin Surface Polymer Layer

Pure atmospheric-pressure hydrogen plasma was employed to etch out the polymer film from the surface of mesh foil and to expose free metal wire surface (Figure 2). The thickness of the polymer layer at the thinnest point over the top of a metal wire was ranging in the order of 1 nm up to 100 nm. The removal of a thin surface polymer film is important to make conductive connection sites on the surface before next processing steps.

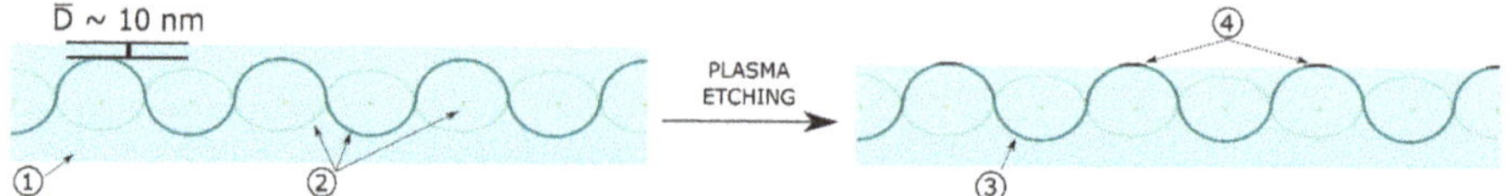

Figure 2. Schematic sketch of plasma etching. The average thickness of the etched polymer over the wire tops was about 10 nm. 1—polymer foil; 2—mesh of polymer and metal wires; 3—metal wire; 4—uncovered metal wires tops.

As seen in Figure 3, already a short plasma exposure time of 1 s can expose "isles" of free metal surface. Figure 3a shows the surface of the untreated sample. More bright, uncovered metal areas are visible in Figure 3b showing etched surface with details on single uncovered wires in Figure 3c,d. Both SEM images Figure 3a,b were taken at the same magnification 26×, working distance 15.0 mm and view field 10.7 mm at accelerating voltage 15.0 kV. The shorter exposure times (0.25 s and 0.5 s) led to the etching of surface polymer film, however, not all metal wires tops were exposed (uncovered) on the surface after such short plasma etchings.

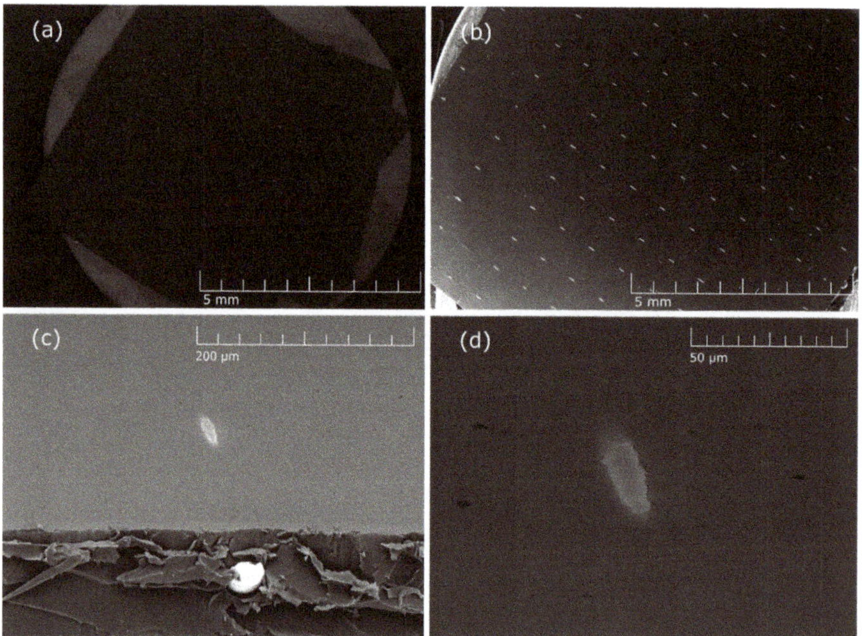

Figure 3. The surface of (**a**) untreated mesh foil and (**b**) the mesh foil after H_2 plasma etching (1 s) with details on cross-section and upper surface of exposed wires (**c**) and upper surface detail (**d**). The distance of the sample surface from the surface of concave DCSBD electrode system was H = 0.35 mm.

Figure 4 shows SEM micrograph of a single uncovered wire after 1 s etching in pure H_2 plasma. The SEM image is shown using secondary electrons (SE image) and also back-scattered electrons (BSE image).

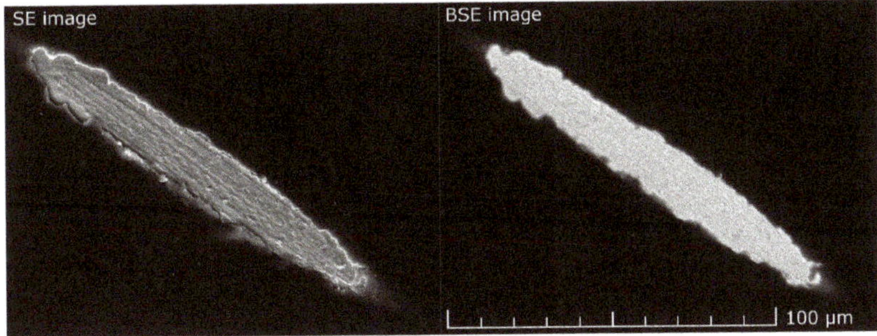

Figure 4. SEM micrograph of the uncovered wire top after 1 s H_2 plasma exposure (distance H = 0.35 mm).

Secondary electrons originate within a few nanometers from the sample surface [20]. Therefore, SE image shows the best the topography of the analyzed surface. Since heavy elements in the specimen reflect or backscatter electrons more effectively than the light elements, the atoms with a high atomic number appear brighter in the BSE image [20]. Backscattered electrons can be then used to detect contrast between areas with different chemical compositions. As seen in Figure 4 (BSE image), we can distinguish a bright uncover metal wire from the dark area composed of light polymer material.

Figure 5a,b shows the EDX line scans in lateral and longitudinal direction, respectively. As observed, surface polymer film was successfully removed uncovering the metal wire. Figure 6 shows composite element map of single wire indicating removal of polymer on the entire area of the uncovered wire. In the area of uncovered wire a strong signal of silver was evident, while only a small signal of copper indicating concentration of copper less than 0.5 wt.% was discernible (Figure 6 and Table 1). This indicates that the thin polymer film was removed from the area above the wire without damage of metal wire and its silver coating. The presence of carbon and a small concentration of oxygen in the area of uncovered wire as shown in Figure 5 can be explained by the surface contamination during the manipulation with the sample, when the sample was exposed to air after the plasma etching for several hours before the SEM EDX analysis. We suppose that this is also the reason for the higher concentration of carbon in the plasma exposed area of pure polymer compared to the value expected on the clean PMMA surface.

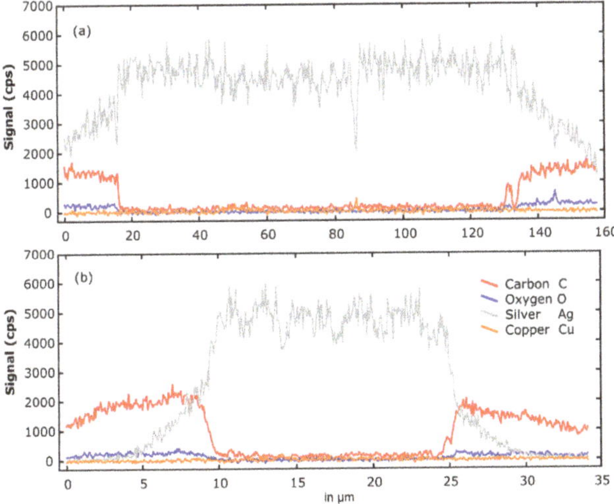

Figure 5. Longitudinal (**a**) and lateral (**b**) EDX line-scans taken on single uncovered wire on sample treated 1 s in H_2 plasma (distance H = 0.35 mm).

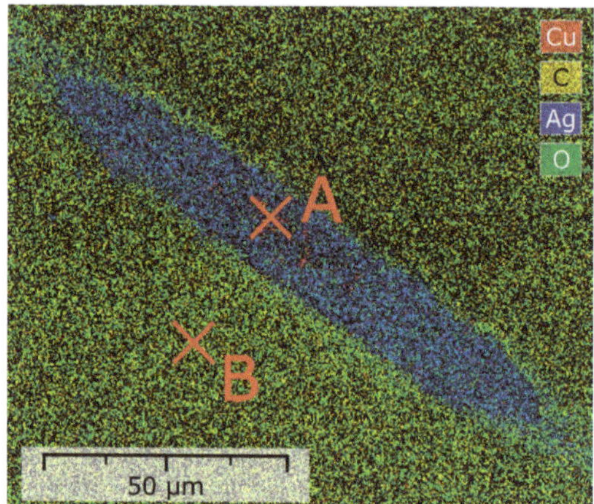

Figure 6. Composite element map in the area of single uncovered wire. C—yellow, O—green, Ag—blue, Cu—red (distance 0.35 mm, treatment time 1 s).

Table 1. Concentration of Carbon, Oxygen, Silver and Copper on the sample after 1 s etching at distance 0.35 mm measured in the areas A and B as shown in Figure 6.

Element		Concentration (at.%)	
		A	B
C	Carbon	34.5	92
O	Oxygen	6.5	8
Ag	Silver	58.5	-
Cu	Copper	0.5	-

3.2. Selectivity of Hydrogen Plasma Etching

Figure 7 shows the details on the tops of uncovered metal wires after 3-s etching in ambient air plasma (a), nitrogen plasma (b), H_2/N_2 plasma (50%) (c), hydrogen plasma (d) and after 1-s etching in hydrogen plasma (e). As seen, ambient air plasma and nitrogen plasma led, besides the polymer etching, also to significant damages of the metal wires. Sample (c) treated in a mixture of hydrogen and nitrogen showed a degradation of polymer around uncovered metal tops. Sample (d) was multiple treated with short exposures of 0.25 s and total treatment time 3.0 s (with a dead time 2.7 s between each plasma exposure and 29.7 s in total). The reason for such non-continuous treatment was to decrease the thermal load on the thermally sensitive PMMA-based substrate and to prevent the unwanted effect of plasma over-exposure. Generally, it is evident that only the pure DCSBD hydrogen plasma has the capability of selective etching of thin surface polymer accompanied by just a negligible effect on the metal wires in the polymer mesh foil.

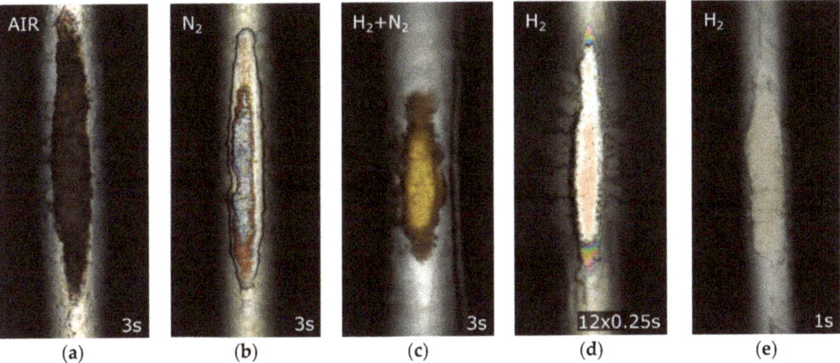

Figure 7. Exposed metal wires after etching in (**a**) ambient air plasma, (**b**) nitrogen plasma, (**c**) H_2/N_2 plasma (50%) and (**d**,**e**) hydrogen plasma. Samples (**a**–**d**): treatment time 3 s (12 × 0.25 s for sample (**d**)); sample (**e**): treatment time 1 s in 100% hydrogen plasma.

3.3. The Effect of the Sample Distance from the DCSBD Plasma Layer on Plasma Etching

Diffuse coplanar surface barrier discharge in a configuration with a concavely curved electrode arrangement is capable of generation of visually almost uniform diffuse plasma layer of some 0.3-mm effective thickness. In this work, the polymer foil was treated at a distance in the range 0.2–0.6 mm from the alumina ceramics.

It was already reported that DCSBD plasma consists of optically two distinctive regions: (i) diffuse plasma located directly above strip electrodes and (ii) the gentle streamer filaments located above the space between the electrodes [21]. It was shown that the decreasing gap between the conductive sample and the ceramics with the plasma layer led to the change of generated plasma— extinction of filamentary plasma was observed.

Figure 8 summarizes the results of 1 s plasma etching in pure H_2 plasma at distances 0.2–0.6 mm. Original SEM images appear in Figure 9. As observed, there is an optimal distance for plasma etching of polymer mesh foil. If the distance is larger than 0.5 mm, the surface of the sample is disrupted by the streamers which are generated perpendicularly to its surface. This results in inhomogeneous etching and uncovering the metal wire only at a small area, as seen in Figure 9d, for the distance of 0.6 mm, together with the rough transition zone between polymeric and metal area. If there was a small gap between the sample and the plasma layer, the streamers could not be creating, and plasma was generated preferably in the diffuse mode. On the other hand, a small gap between the sample and the ceramics with the electrode system led to an increase of thermal heating of the polymer. As seen from SEM micrograph in Figure 9a, thermal heating may cause a cracking of the polymer in the transition zone around a free metal wire. A roughening of the wire and sputtering of the thin silver coating from the wire was observed.

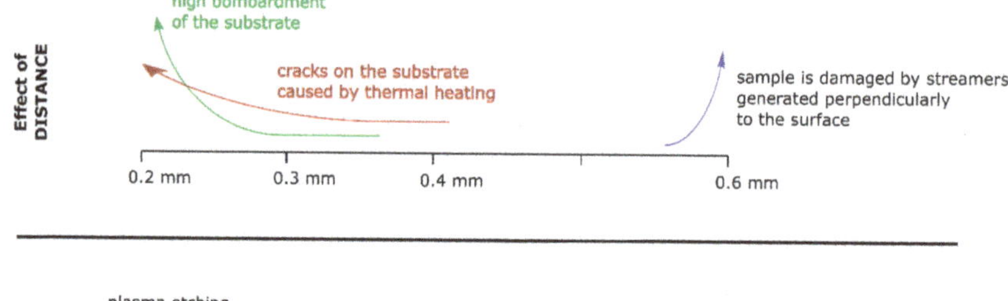

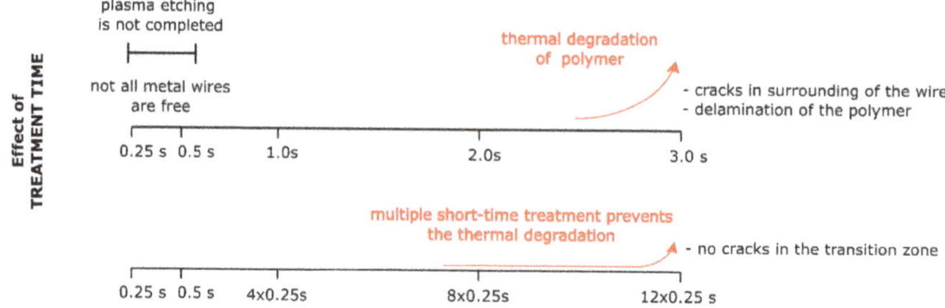

Figure 8. The effect of the sample distance from DCSBD plasma layer, and H_2 plasma treatment time on the mechanism of the etching of polymer mesh foil.

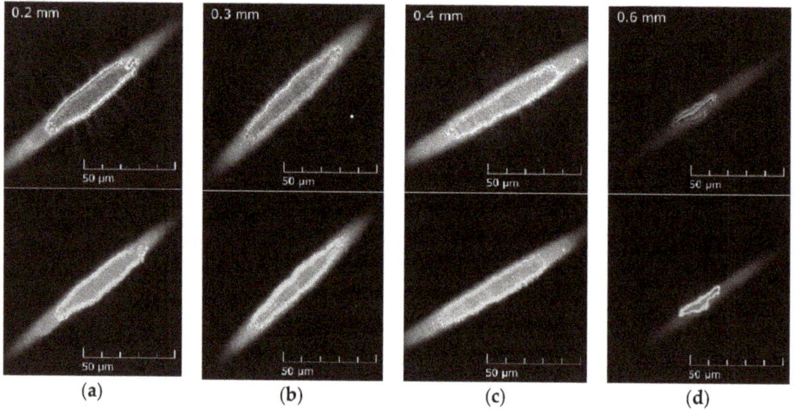

Figure 9. SEM micrographs of free metal wires after 1 s plasma etching in pure H_2 plasma at distance of 0.2 mm (**a**), 0.3 mm (**b**), 0.4 mm (**c**) and 0.6 mm (**d**). Top row—SE image; bottom row—BSE image.

3.4. The Effect of the Treatment Time on Plasma Etching

As reported, only 1 s exposure to H_2 plasma can etch the surface of polymer film sufficiently to uncover of the metal wires tops. The exposures less than 0.5 s resulted in not completed etching when not all metal wires tops were completely free of the polymer film. The longer etching in hydrogen DCSBD plasma resulted in the cracks in the polymer around the uncovered wires tops and delamination of the polymer as seen in Figure 10. This damage was caused by thermal degradation of the polymer at longer exposures in plasma. The benefit of using 100% hydrogen plasma treatment is also in a significantly lower thermal impact on the surface compared to 100% nitrogen atmospheric plasma or H_2/N_2 gas mixture due to the absence of any heavy particle. A surface power density of 100% hydrogen plasma was about 2.5 W cm^{-2}, whereas the electron density and electron temperature were 1.3×10^{16} cm^{-3} and 19×10^3 K respectively [15].

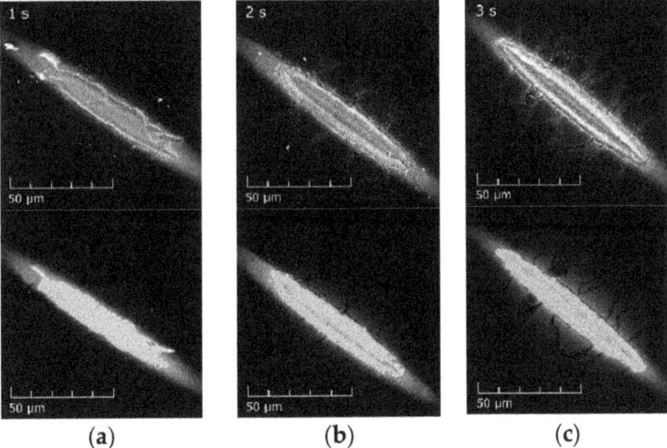

Figure 10. SEM micrographs of free metal wires treated for (**a**) 1 s (**b**), 2 s (**c**) and 3 s at distance H = 0.3 mm. Top row—SE image; bottom row—BSE image.

It was found that the unwanted thermal degradation of polymer foil can be prevented by applying a multiple short-time plasma exposure on the etched surface. This is apparent from Figure 11 comparing the samples etched for 3 s using single 3 s plasma exposure (Figure 11a) and the multiple very short exposures (12 × 0.25 s) in hydrogen plasma (Figure 11b).

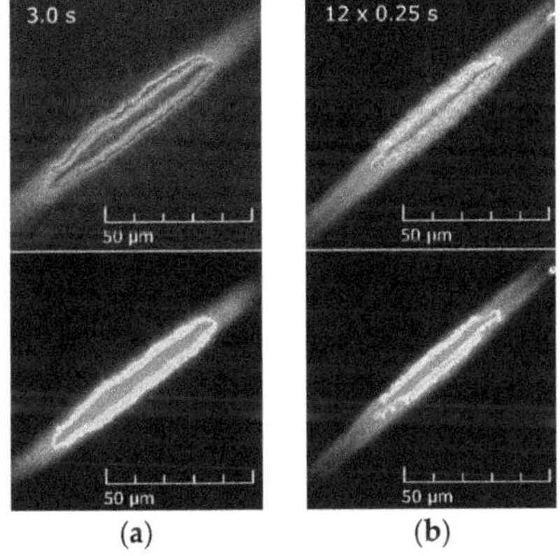

Figure 11. SEM micrographs of metal wires after 3 s plasma etching in pure H_2 plasma at distance H = 0.6 mm for single 3 s plasma exposure (**a**) and the multiple very short exposures (12 × 0.25 s) (**b**). Top row—SE image; bottom row—BSE image.

4. Discussion

As confirmed by SEM/EDX analysis, the atmospheric pressure DCSBD plasmas generated in ambient air, pure nitrogen, pure hydrogen and in mixtures of nitrogen with hydrogen is capable of etching of PMMA polymer layers on polymer-metal mesh substrates. However, a strong degradation of the polymer substrate and also the significant damages to the metal wires were observed. On the other hand, it was found that using a short, 1 s exposure of the polymer surface to the pure hydrogen plasma it is possible to etch selectively the surface polymer layer and to uncover the metal wires without damaging them. As a consequence, we conclude that the pure hydrogen DCSBD plasma can be used for the fast, selective etching of the polymer film surface without any destruction of the metal wires. We observed that the etching mechanism strongly depends on the parameters such as treatment time and the distance of the sample from discharge. Therefore, the effect of these parameters was discussed in more details.

The application of pure hydrogen plasma raises the questions about the safety of the process. As in any other activity/process, it is necessary to find a trade-off between the safety, the process performance, and the product requirements. The presented plasma modification method might be improved by localized feeding of working gas just into the very narrow region where the plasma is generated. Such technical solution would enable to decrease the working gas consumption and together with smaller volume of (dangerous) working gas presented in the reactor chamber this would lead to higher safety of the entire process and whole device.

In comparison to standard mechanical abrasion and low-pressure plasma etching of studied flexible polymer-metal mesh substrate, the presented method utilizing atmospheric DCSBD plasma is easily scalable and compatible with roll-to-roll processing of large-area substrates. To sum up, we conclude that the DCSBD plasma sources can be used for fast, large-area, roll-to-roll, selective plasma etching of complex materials like flexible transparent conductive substrates which can be applied e.g., for flexible photovoltaic applications and other emerging technologies.

Author Contributions: Conceptualization, M.Č. and D.K.; investigation, R.K., J.J., M.Z. and J.K.; writing—original draft preparation, R.K.; writing—review and editing, R.K., D.K. and M.Č. All authors have read and agreed to the published version of the manuscript.

Funding: This research was supported by project LM2018097 funded by the Ministry of Education, Youth and Sports of the Czech Republic and project TJ01000327 funded by the Technology Agency of the Czech Republic.

Acknowledgments: The authors thank to Peter Chabrecek and SEFAR AG (Switzerland) for providing the polymer mesh substrate. Thanks belongs also to Jan Čech and Dana Skácelová both from CEPLANT, MU Brno for carrying out the trial experiments on selective etching of small polymer mesh samples providing by SEFAR AG.

Conflicts of Interest: The authors declare no conflict of interest.

References

1. Puliyalil, H.; Cvelbar, U. Selective plasma etching of polymeric substrates for advanced Applications. *Nanomaterial* **2016**, *6*, 108. [CrossRef] [PubMed]

2. Sarra-Bournet, C.; Ayotte, G.; Turgeon, S.; Massines, F.; Laroche, G. Effects of chemical composition and the addition of H_2 in a N_2 atmospheric pressure dielectric barrier discharge on polymer surface functionalization. *Langmuir* **2009**, *25*, 9432–9440. [CrossRef] [PubMed]

3. Vargo, T.G.; Gardella, J.A.; Meyer, A.E.; Baier, R.E. Hydrogen/liquid vapor radio frequency glow discharge plasma oxidation/hydrolysis of expanded poly(tetrafluoroethylene) (ePTFE) and poly (vinylidene fluoride) (PVDF) surfaces. *J. Polym. Sci. Part A Polym. Chem.* **1991**, *29*, 555–570. [CrossRef]

4. Cunge, G.; Ferrah, D.; Petit-Etienne, C.; Davydova, A.; Okuno, H.; Kalita, D.; Bouchiat, V.; Renault, O. Dry efficient cleaning of poly-methyl-methacrylate residues from graphene with high-density H_2 and H_2-N_2 plasmas. *J. Appl. Phys.* **2015**, *118*, 123302. [CrossRef]

5. Xie, L.; Jiao, L.; Dai, H. Selective etching of graphene edges by hydrogen plasma. *J. Am. Chem. Soc.* **2010**, *132*, 14751–14753. [CrossRef] [PubMed]

6. Hou, P.-X.; Li, W.-S.; Zhao, S.; Li, G.-X.; Shi, C.; Liu, C.; Cheng, H.-M. Preparation of metallic single-wall carbon nanotubes by selective etching. *ACS Nano* **2014**, *8*, 7156–7162. [CrossRef] [PubMed]

7. Zhang, G.; Qi, P.; Wang, X.; Lu, Y.; Li, X.; Tu, R.; Bangsaruntip, S.; Mann, D.; Zhang, L.; Dai, H. Selective etching of metallic carbon nanotubes by gas-phase reaction. *Science* **2006**, *314*, 974–977. [CrossRef] [PubMed]

8. Šimor, M.; Ráheľ, J.; Vojtek, P.; Černák, M.; Brablec, A. Atmospheric-pressure diffuse coplanar surface discharge for surface treatments. *Appl. Phys. Lett.* **2002**, *81*, 2716–2718. [CrossRef]

9. Georghiou, G.E.; Papadakis, A.P.; Morrow, R.; Metaxas, A.C. Numerical modelling of atmospheric pressure gas discharges leading to plasma production. *J. Phys. D Appl. Phys.* **2005**, *38*, R303–R328. [CrossRef]

10. Roth, J.R. *Industrial Plasma Engineering: Volume 2: Applications to Nonthermal Plasma Processing*, 1st ed.; CRC Press: Boca Raton, FL, USA, 2001.

11. Kuzminova, A.; Kretková, T.; Kylián, O.; Hanuš, J.; Khalakhan, I.; Prukner, V.; Doležalová, E.; Šimek, M.; Biederman, H. Etching of polymers, proteins and bacterial spores by atmospheric pressure DBD plasma in air. *J. Phys. D Appl. Phys.* **2017**, *50*, 135201. [CrossRef]

12. Kwon, Y.-T.; Lee, Y.-I.; Kim, S.; Lee, K.-J.; Choa, Y.-H. Full densification of inkjet-printed copper conductive tracks on a flexible substrate utilizing a hydrogen plasma sintering. *Appl. Surf. Sci.* **2017**, *396*, 1239–1244. [CrossRef]

13. Černák, M.; Kováčik, D.; Ráheľ, J.; Sťahel, P.; Zahoranová, A.; Kubincová, J.; Tóth, A.; Černáková, Ľ. Generation of a high-density highly non-equilibrium air plasma for high-speed large-area flat surface processing. *Plasma Phys. Control Fusion* **2011**, *53*, 124031. [CrossRef]

14. Homola, T.; Pospisil, J.; Krumpolec, R.; Souček, P.; Dzik, P.; Weiter, M.; Černák, M. Atmospheric Dry Hydrogen Plasma Reduction of Inkjet-Printed Flexible Graphene Oxide Electrodes. *ChemSusChem.* **2018**, *11*, 941–947. [CrossRef] [PubMed]

15. Prysiazhnyi, V.; Brablec, A.; Čech, J.; Stupavská, M.; Černák, M. Generation of Large-Area Highly-Nonequlibrium Plasma in Pure Hydrogen at Atmospheric Pressure. *Contrib. Plasma Phys.* **2014**, *54*, 138–144. [CrossRef]

16. Krumpolec, R.; Čech, J.; Jurmanová, J.; Ďurina, P.; Černák, M. Atmospheric pressure plasma etching of silicon dioxide using diffuse coplanar surface barrier discharge generated in pure hydrogen. *Surf. Coatings Technol.* **2017**, *309*, 301–308. [CrossRef]

17. Jirásek, V.; Čech, J.; Kozak, H.; Artemenko, A.; Černák, M.; Kromka, A. Plasma treatment of detonation and HPHT nanodiamonds in diffuse coplanar surface barrier discharge in H_2/N_2 flow. *Phys. Status Solidi (A)* **2016**, *213*, 2680–2686. [CrossRef]

18. Tučeková, Z.; Kelar, J.; Doubková, Z.; Krumpolec, R. Structuring of polymethylmethacrylate substrates by reducing plasma. In *Book of Contribution Papers, Proceedings of the 22nd Symp. Appl. Plasma Process. (SAPP XXII)* 11th EU-Japan Jt. Symp. Plasma Process, Štrbské Pleso, Slovakia, 18–24 January 2019; Medvecká, V., Országh, J., Papp, P., Matejčík, Š., Eds.; Society for Plasma Research and Applications in Cooperation with Library and Publishing Centre CU: Bratislava, Slovakia, 2019; pp. 269–272.

19. Doubková, Z.; Tučeková, Z.; Kelar, J.; Krumpolec, R.; Zemánek, M. Modification of various polymer surfaces using atmospheric pressure reducing plasma. In Proceedings of the NANOCON 2018, Brno, Czech Republic, 17–19 October 2018; Peer Rev. TANGER Ltd.: Ostrava, Czech Republic, 2019; pp. 688–693. Available online: https://www.nanocon.eu/files/uploads/01/NANOCON2018-ConferenceProceedings_content.pdf (accessed on 15 October 2020).

20. Egerton, R.F. *Physical Principles of Electron Microscopy*; Springer-Verlag US: Boston, MA, USA, 2005; ISBN 978-0-387-25800-3.

21. Homola, T.; Matoušek, J.; Medvecká, V.; Zahoranová, A.; Kormunda, M.; Kováčik, D.; Černák, M. Atmospheric pressure diffuse plasma in ambient air for ITO surface cleaning. *Appl. Surf. Sci.* **2012**, *258*, 7135–7139. [CrossRef]

Publisher's Note: MDPI stays neutral with regard to jurisdictional claims in published maps and institutional affiliations.

Article

Study on the Effect of Structure Parameters on NO Oxidation in DBD Reactor under Oxygen-Enriched Condition

Yunkai Cai *, Lin Lu and Peng Li

Marine Engineering, School of Energy and Power Engineering, Wuhan University of Technology, Wuhan 430070, China; lulinwhut@whut.edu.cn (L.L.); leepeng@whut.edu.cn (P.L.)
* Correspondence: caiyunkai@whut.edu.cn; Tel.: +86-1562-372-3935

Received: 4 September 2020; Accepted: 24 September 2020; Published: 27 September 2020

Abstract: To improve NO oxidation and energy efficiency, the effect of dielectric barrier discharge reactor structure on NO oxidation was studied experimentally in simulated diesel exhaust at atmospheric pressure. The mixture of 15% O_2/N_2 (balance)/860 ppm NO_X (92% NO + 8% NO_2) was used as simulated diesel exhaust. The results show that DBD reactor with 100-mm electrode length has the highest oxidation degree of NO_X and energy efficiency. NO oxidation efficiency is promoted and the generation of NO is inhibited significantly by increasing the inner electrode diameter. Increasing the inner electrode diameter not only improve the E/N, but also makes the distribution of E/N more concentrated in the gas gap. The secondary electron emission coefficient (γ) of electrode material is closely related to electron energy and cannot be considered as a constant, which causes the different performance of electrode material for NO oxidation under different gas gap conditions. Compared with the rod electrode, the screw electrode has a higher electric field strength near the top of the screw, which promotes the generation of N radicals and inhibits the generation of O radicals. Rod electrode has a higher NO oxidation and energy efficiency than screw electrode under oxygen-enriched condition.

Keywords: dielectric barrier discharge; NO oxidation; diesel exhaust; oxidation degree of NO_X

1. Introduction

Ocean transportation undertakes more than 90% of the global freight transportation which has the advantages of low cost and high security [1]. Furthermore, ships generally use diesel engines as their main propulsion devices. Sometimes, high-sulfur oil is used as the alternative fuel in marine diesel engines for lower shipping costs. As a result, their exhaust emissions cause serious air pollutions [2–4]. To control the exhaust pollution of marine diesel engine, the International Maritime Organization (IMO) regulated the emission of NO_X and SO_X from marine diesel engine exhaust, as shown in Figure 1 [5,6].

For SO_X removal, wet scrubbing can achieve a higher removal efficiency (more than 95%), while, for NO_X removal, wet scrubbing has a low removal efficiency. This because more than 90% of NO_X in the exhaust gas of marine diesel engine is NO, and NO has low solubility in water and does not react with alkali solution. Some studies have shown that NO_2 can promote the absorption of NO by alkaline solution. When the oxidation degree of NO_X is increased to ~50%, the absorption efficiency of NO_X by alkaline solution is the highest [7]. The technology of wet desulfurization and denitrification is easy to integrate with existing desulfurization scrubber, reducing the investment and space, which will be an important development trend in the future [4]. Therefore, non-thermal plasma (NTP) oxidation combined with wet scrubbing for simultaneous desulfurization and denitrification was widely studied [8–10]. Compared to electron beam (EB) and corona discharge (CD), dielectric barrier discharge (DBD) has higher NO removal efficiency at lower specific energy density (SED) and shorter residence time [11]. There are many types of DBD, such as volume dielectric barrier discharge (VDBD), surface dielectric barrier discharge (SDBD) and diffuse coplanar surface barrier discharge (DCSBD) [12,13]. SDBD and DCSBD were widely used for surface modification. The VDBD reactor was widely used for the treatment of NO_X and SO_X. The structure parameters of DBD reactor have great impact on NO removal efficiency and energy consumption, which has become a research hotspot [14–19].

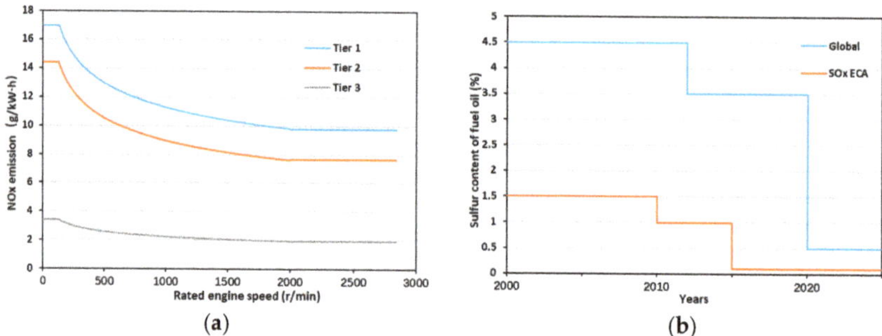

Figure 1. (**a**) Regulation of NO_X emission; and (**b**) regulation of SO_X emission. ECA, emission control area. Tier 1 took effect in 2000, Tier 2 in 2011 and Tier 3 in 2016.

In 2011 and 2012, Wanget al. studied the effect of inner electrode shape on discharge of DBD reactor. They showed that the DBD reactor with multineedle-to-cylinder electrode had better discharge performance than the regular DBD reactor [14,15]. In 2012 and 2013, Wang et al. used the mixture of NO and N_2 to simulate the exhaust gas. The effects of electrode shape, electrode material and barrier medium thickness on NO removal were studied. They showed that tungsten electrode has a higher NO removal efficiency than copper and stainless-steel electrode with the same SED, which has a higher secondary electron emission coefficient. Screw electrode has a higher NO removal efficiency than rod electrodes, due to the smaller air gap capacitance, which makes the driving power of DBD reactor lower and dielectric loss lower [16,17]. In 2015 and 2016, Anaghizi and Talebizadeh et al. used the mixture of NO and N_2 as the simulated diesel exhaust. The effects of electrode length, electrode material and electrode diameter on NO removal were studied [18,19].

Although the structure of DBD reactor has been studied in many previous investigations, most denitrification was carried out under the condition of NO_X/N_2. In such circumstance, NO_X is mainly removed by reduction reaction. However, the real exhaust gas contains a large amount of O_2, which makes the reduction effect of NO_X by collision with N radicals ignorable. NO_X mainly reacts with O, OH, HO_2 and other radicals produced by O_2 and H_2O and needs wet scrubber for complete removal [20]. There are few studies on the effect of DBD reactor structure parameters on NO oxidation under oxygen-enriched condition. Therefore, this study used 860 ppm NO_X (92% NO + 8% NO_2)/15% O_2/N_2 (balance) mixture as the simulated diesel exhaust gas. The effects of DBD structure parameters (such as electrode length, inner electrode diameter, inner electrode material and inner electrode shape) on NO oxidation performance were investigated. Supply voltage and discharge power were measured as well.

2. Experiment and Methods

2.1. Experiment Setup

The schematic diagram of the experimental system is shown in Figure 2. The gas flow of each component is controlled by the mass flow controller (MFC) and mixed uniformly by the gas mixer before entering the DBD reactor. The exhaust gas after DBD treatment is absorbed by alkali liquor, which reduces the damage to the environment. The concentrations of NO, NO_2 and O_2 were measured by flue gas analyzer (Horiba PG-350). The total gas flow rate is 5 L/min and the gas temperature is 25 °C.

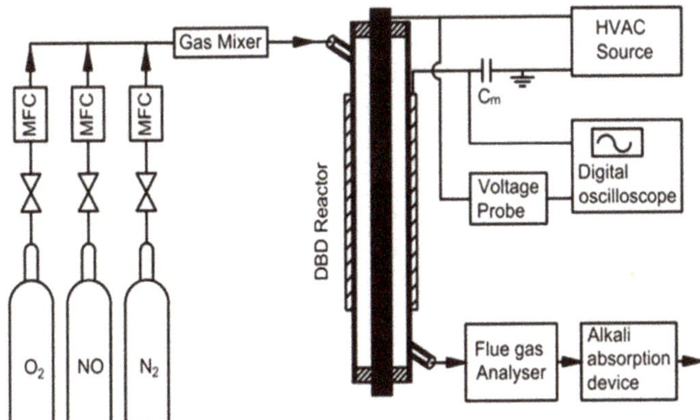

Figure 2. The schematic diagram of the experimental setup.

The high voltage AC (HVAC) power supply is used as the power supply of DBD reactor. The supply voltage used to generate plasma in the DBD reactor is within 0–25 kV and discharge frequency is 10 kHz. The discharge power of DBD reactor is controlled by changing the supply voltage. The measurement capacitance (C_m) of DBD discharge transmission charge is 0.47 µF, and the discharge voltage (voltage between inner and outer electrode) is decreased 1000 times by high voltage probe (TEK P6015A), which is input into oscilloscope. The discharge power is calculated by Lissajous figure. Figure 3 shows Lissajous figure obtained by digital oscilloscope (TEK TBS1052b).

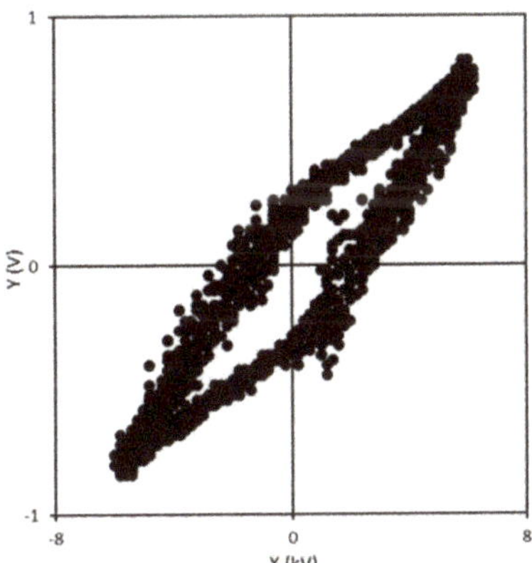

Figure 3. Lissajous figure obtained by digital oscilloscope (supply voltage V_{PP} =14 kV, DBD reactor with inner electrode diameter 14 mm, electrode length 100 mm).

A cross-sectional view of DBD reactor and inner electrodes are shown in Figure 4. The DBD reactor is coaxial cylindrical, and the dielectric barrier is a quartz tube. The length of the quartz tube is 450 mm, the outer diameter is 18 mm, the inner diameter is 15 mm and the thickness is 1.5 mm. The quartz tube is wrapped with copper foil as the outer electrode, and the length of discharge interval is adjusted by changing the length of copper foil. The materials making up the inner electrode are copper, stainless steel and aluminum with diameters of 10, 12 and 14 mm. The shape of the inner electrode is rod or screw.

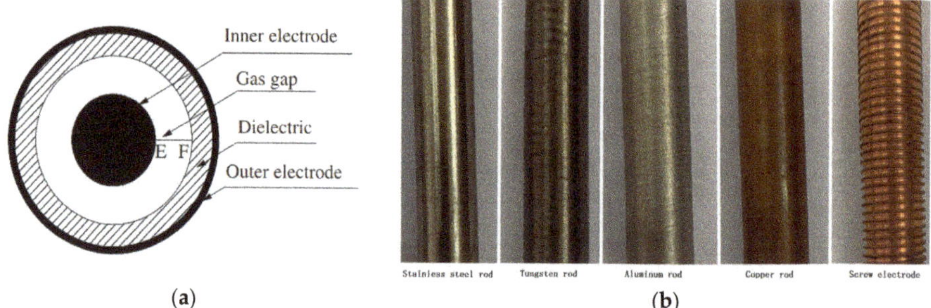

(a) (b)

Figure 4. (**a**) A cross-sectional view of the DBD reactor; and (**b**) the various inner electrodes used in the experiment.

2.2. Evaluation Methods of NO Oxidation and Energy Efficiency

In each test, the concentration of NO, NO_2 and NO_X (NO+NO$_2$) were measured by gas analyzer for 30 times, as shown in Figure 5. The standard deviations of NO, NO_2 and NO_X are 2.2, 3.3 and 1.8 ppm, respectively. The average concentrations of NO, NO_2 and NOx were taken as the final test results.

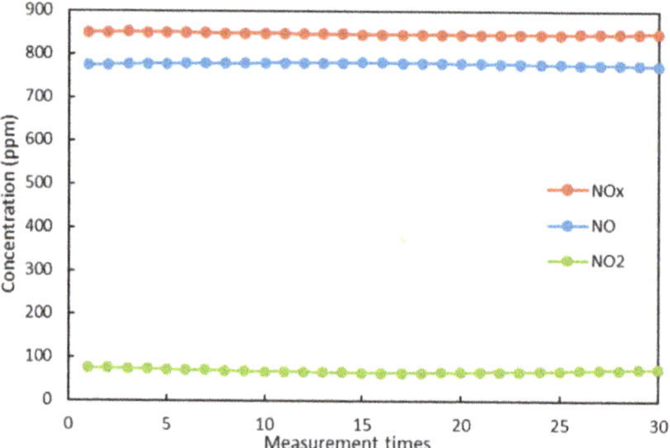

Figure 5. The concentration of NO, NO_2 and NO_X (NO+NO$_2$) measured by gas analyzer.

To explore the effect of DBD structure on NO oxidation, the oxidation degree of NO_X was selected as the evaluation index. The oxidation degree of NO_X is defined as follows:

$$\alpha = \frac{\varphi(NO_2)}{\varphi(NO_x)} \times 100\% \tag{1}$$

where α is oxidation degree of NO_X; $\varphi(NO_2)$ is the concentration of NO_2, ppm; and $\varphi(NO_x)$ is the concentration of NO_X, ppm. The discharge power was obtained by Lissajous figure (see Figure 3). The calculation of discharge power is given in Equation (2):

$$P = fC_mK_xK_yKA, \tag{2}$$

where f is discharge frequency, Hz; K_x is x-axis sensitivity of oscilloscope, V/grid; K_y is y-axis sensitivity of oscilloscope, V/grid; K is voltage decrease ratio of high voltage probe, 1000:1; and A is area enclosed by Lissajous figure. To evaluate the energy consumption level of DBD reactor, the specific energy density (SED) is defined as follows:

$$SED = \frac{P}{Q} \times 60,\tag{3}$$

where SED is specific energy density, J/L; P is discharge power, W; and Q is gas flow, L/min.

3. Results and Discussion

3.1. Effect of Electrode Length on NO Oxidation

Figure 6 shows the effect of electrode length on NO oxidation under different SED. The experimental results show that the 100- and 150-mm electrodes have a higher oxidation degree of NO_X than 50-mm electrode. The energy consumption is lower when the electrode length is 100 mm. Electrode length determines the length of discharge interval, which mainly affects the residence time of gas in the discharge interval of DBD reactor [18,21]. Plasma chemical reaction is mainly divided into two stages. In the primary stage, the electrons are accelerated to form high-energy electrons by electric field in the discharge range, and the high-energy electrons collide with gas molecules leading to the ionization, excitation and dissociation of neutral molecules. In the secondary stage, free radicals, excited-state molecules and ions interact to cause radical–radical, radical–neutrals and ion–ion reactions in the downstream of the discharge range [22]. Therefore, the free radicals, excited molecules and atoms generated in the discharge range have a significant impact on pollutant removal. Under oxygen-enriched condition, the concentration of oxidative radicals in the discharge range of DBD reactor gradually increased and reached equilibrium within several discharge cycles, and then it was basically stable with time [23]. It can be seen from the experimental results in Figure 6 that, when the electrode length is 50 mm, the maximum oxidation degree of NO_X is lower than that of electrode lengths 100 and 150 mm due to the short residence time of gas in DBD reactor, and the oxidation radicals do not reach the equilibrium concentration. When the electrode length is in the range of 100–150 mm, the concentration of oxidation free radicals reaches equilibrium and does not change with the increase of discharge time. Increasing the electrode length does not lead to higher NO_X oxidation degree, but higher energy consumption.

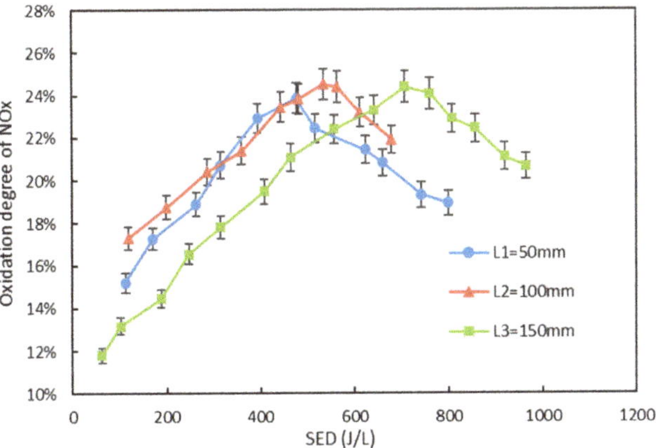

Figure 6. Effect of electrode length on oxidation degree of NO_X.

It is worth noticing that the oxidation degree of NO_X rises first and then shows a decline trend with the increase of SED under oxygen-enriched condition, which is different from the NO_X removal efficiency behavior under the condition of non-oxygen [19].

To analyze the mechanism, we measured the concentration changes of NO and NO_2 in the process of NO oxidation in DBD reactor under oxygen-enriched condition. Figure 7 shows that the concentration of NO decreases first and then increases, while the concentration of NO_2 increases first and then decreases with the increase of SED. The main plasma chemical reactions involved in oxygen enriched conditions are listed in Table 1.

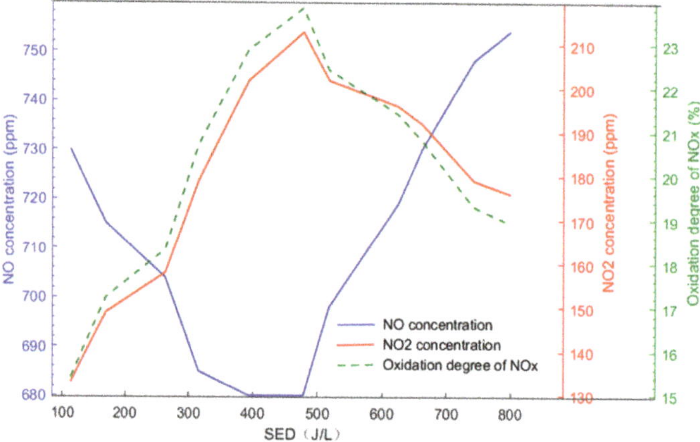

Figure 7. Change of NO and NO_2 concentration during discharge under oxygen-enriched condition (electrode length 50 mm).

Table 1. Main reactions and their rate coefficients.

Reactions	NO.	Rate Coefficients (cm^3/s or cm^6/s)	References
$e + N_2(X) \rightarrow e + N_2(X, v)$	R1	$f(E/N)$	[24]
$e + N_2(X) \rightarrow e + N_2\left(A^3\Sigma_u^+\right)$	R2	$f(E/N)$	[25]
$e + N_2(X) \rightarrow e + N + N$	R3	$f(E/N)$	[25]
$e + O_2 \rightarrow e + O + O$	R4	$f(E/N)$	[26]
$e + O_2 \rightarrow e + O + O(^1D)$	R5	$f(E/N)$	[26]
$O + O_2 + N_2 \rightarrow O_3 + N_2$	R6	$k = 6.2 \times 10^{-34} \times \left(\frac{300}{T}\right)^2$	[27]
$O + O_2 + O_2 \rightarrow O_3 + O_2$	R7	$k = 6.9 \times 10^{-34} \times \left(\frac{300}{T}\right)^{1.25}$	[27]
$O_3 + NO \rightarrow NO_2 + O_2$	R8	$k = 1.4 \times 10^{-12} \times \exp\left(-\frac{1310}{T}\right)$	[28]
$O + NO + M \rightarrow NO_2 + M$	R9	$k = 1.03 \times 10^{-30}\left(\frac{T}{298}\right)^{-2.87} \exp\left(\frac{780}{T}\right)$	[28]
$N_2(X, v \geq 13) + O \rightarrow NO + N$	R10	$k = 1 \times 10^{-13}$	[29]
$N + O_2 \rightarrow NO + O$	R11	$k = 1.1 \times 10^{-14} \times T \times \exp\left(-\frac{3150}{T}\right)$	[29]
$N_2\left(A^3\Sigma_u^+\right) + O_2 \rightarrow N_2 + O + O$	R12	$k = 1.63 \times 10^{-12} \times \left(\frac{T}{300}\right)^{0.55}$	[29]
$N_2\left(A^3\Sigma_u^+\right) + O \rightarrow NO + N$	R13	$k = 7 \times 10^{-12}$	[29]
$N + O + N_2 \rightarrow NO + N_2$	R14	$k = 1.76 \times 10^{-31} \times T^{-0.5}$	[29]
$N + O_3 \rightarrow NO + O_2$	R15	$k = 5 \times 10^{-16}$	[29]

The discharge voltage increases with the increase of SED, as shown in Figure 8a. As a result, the reduced electric field strength E/N (ratio of the electric field strength to the gas particle number density) is increased (as shown in Figure 8b), which makes the mean electron energy increase [16].

Because the dissociation energy of N_2 (9.8 eV) is higher than that of O_2 (5.1 eV), the formation of O radicals is easier than that of N radicals in the process of plasma discharge [30]. O and O_3 are generated firstly in the DBD reactor, which promote the oxidation of NO to NO_2 via R4–R9 [26–28]. Therefore, the concentration of NO decreases, the concentration of NO_2 increases and the oxidation degree of NO_X increases gradually with the increase of SED at the initial discharge stage.

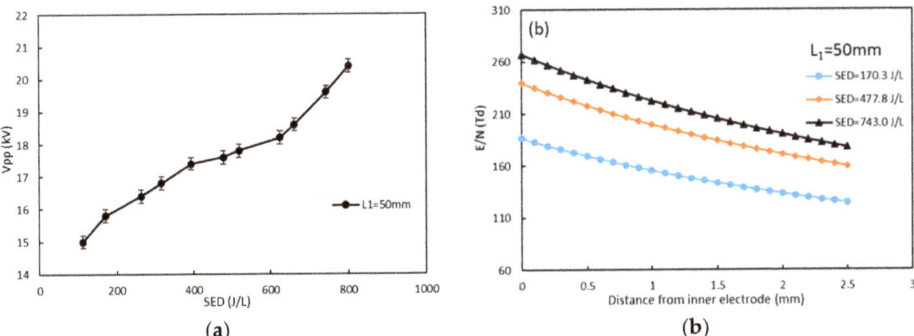

Figure 8. (**a**) The effect of SED on discharge voltage; and (**b**) the effect of SED on E/N in gas gap of DBD reactor (electrode length is 50 mm and V_{pp} means peak-to-peak voltage).

With the increase of SED, the E/N further increases, which promotes the dissociation of N_2 to generate N radicals via R3, especially when the E/N accesses 200 Td [16]. Under oxygen-enriched condition, the N radicals will react with O, O_2 and O_3 to generate NO via R11, R14 and R15 [29], which causes the increase of NO.

To analyze the effect of E/N on the generation of radicals and excitation species in NTP process, BOLSIG+ were used to calculate the electron energy distribution function (EEDF) and mean electron energy under different E/N conditions. The electron Boltzmann equation (BE) in an ionized gas is [31]:

$$\frac{\partial f}{\partial t} + v \cdot \nabla f - \frac{e}{m} E \cdot \nabla_v f = C[f] \tag{4}$$

where f is the electron distribution in six-dimensional phase space, v are the velocity coordinates, e is the elementary charge, m is the electron mass, E is the electric field, ∇_V is the velocity-gradient operator and C represents the rate of change in f due to collisions.

The mean electron energy was given by:

$$\bar{\varepsilon} = \int_0^\infty \varepsilon^{3/2} f_0 d\varepsilon \tag{5}$$

where $\bar{\varepsilon}$ is the mean electron energy, f_0 is isotropic part of EEDF.

The rate coefficient related to electron collisions is defined as:

$$k_k = \sqrt{\frac{2e}{m}} \int_0^\infty \varepsilon \sigma_k f_0 d\varepsilon \tag{6}$$

where σ_k is the cross section of collision process k. The electron–molecule collision cross sections are derived from LXCat [24–26].

The efficiency for a particular electron collision process can be calculated by [32]:

$$G\ value = 100k/(v_d E/N) \tag{7}$$

where k is the rate coefficient of a particular electron collision process and v_d is the drift velocity of electron. The G value represents the number of reactions per 100 eV of input energy. The rate coefficient k represents the number of reactions in a unit volume per unit time. The quantity $v_d E/N$ represents the amount of energy expended by the electrons in a unit volume per unit time.

The drift velocity v_d can be calculated from the solution f_0 by Equation (8) [33]:

$$v_d = -\frac{E}{3N}\sqrt{\frac{2e}{m}}\int_0^\infty \frac{\varepsilon}{Q}\frac{\partial f_0}{\partial \varepsilon}d\varepsilon \tag{8}$$

where Q is the effective total momentum-transfer cross section.

The simulation results are shown in Figure 9. Figure 9a shows the effect of E/N on mean electron energy. Obviously, the mean electron energy increases with the increase of E/N.

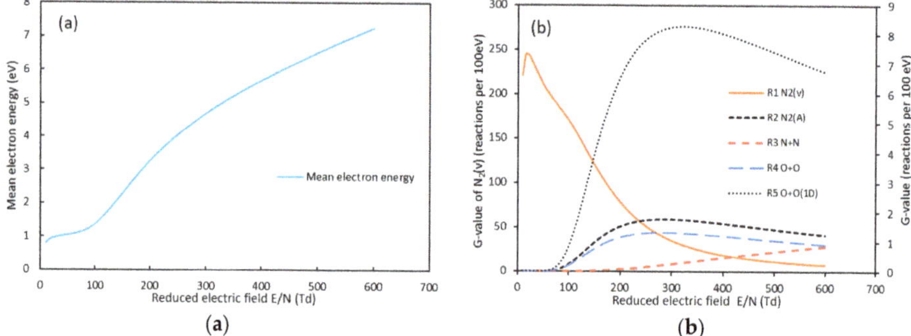

Figure 9. (**a**) The effect of E/N on mean electron energy; and (**b**) the effect of E/N on G-value of particularly reaction process (The G-value of reaction R1 is drawn separately on the left vertical axis).

Figure 9b shows the effect of the E/N on the efficiency for a particular electron collision process. The results show that the G-value of the reaction R1 for the generation of $N_2(X, v)$ (which has been proven to be an important species for the generation of NO via R10 and R11 [34]) decreases rapidly with the increase of E/N. The G-values of the main reactions R4 and R5 for the generation of O radicals increase rapidly in the range of 100–250 Td. As a result, NO is transformed to NO_2 via R8 and R9, which causes the concentration of NO_2 increases, and the oxidation degree of NO_X increases with the increase of SED at the initial discharge stage (as shown in Figures 6 and 7).

The G-value of the reaction R3 for the generation of N radicals increases rapidly when E/N exceeds 250 Td. Although the G-value of the reaction R2 for the formation of $N_2(A^3\Sigma_u^+)$ (which was considered as an important species to generate NO via R13 [35]) decreases when E/N exceeds 250 Td, the generation efficiency of NO is higher due to the higher reaction rate coefficient of N for NO generation than that of $N_2(A^3\Sigma_u^+)$ [30,35,36]. As a result, the generation efficiency of O radical decreases and the generation of NO increases when the E/N exceeds 250 Td, which causes the oxidation degree of NO_X to decrease with the increase of SED (as shown in Figures 6 and 7).

3.2. Effect of Electrode Diameter on NO Oxidation

Figure 10 shows the effect of electrode diameter on NO oxidation under different SED conditions. The experimental results show that increasing the electrode diameter can obtain higher oxidation degree of NO_X and reduce the energy consumption of DBD reactor. Increasing inner electrode diameter can reduce the gas gap of DBD reactor, which promotes the E/N of gas gap with the same supply voltage [16,19]. As E/N increases, the mean electron energy increases as well; thus, increasing the inner electrode diameter makes it easier for DBD reactor to generate O radicals and promote NO oxidation.

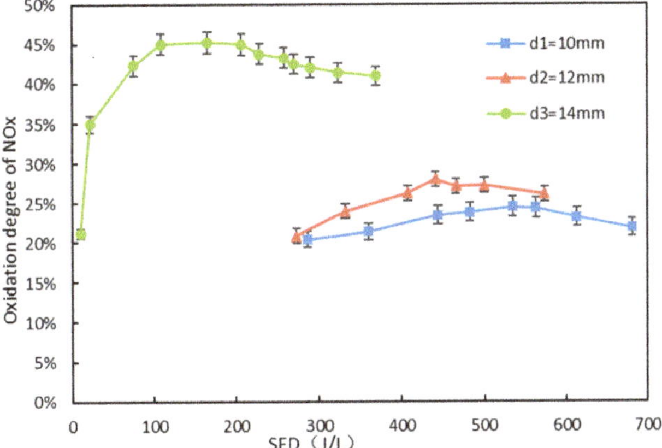

Figure 10. Effect of electrode diameter on oxidation degree of NO_X (electrode length is 100 mm; inner electrode diameter is 10, 12 and 14 mm; and gas gap is 2.5, 1.5 and 0.5 mm, respectively).

Figure 11 shows the relationship between SED and discharge voltage for different electrode diameters. The results show that the discharge voltage decreases with the increase of inner electrode diameter under the same SED. The breakdown voltage (U_b) of coaxial cylindrical DBD reactor can be calculated by the following formula [37]:

$$U_b = \frac{BPd}{ln\left(\frac{APd}{1+\frac{1}{\gamma}}\right)} \tag{9}$$

where U_b is the gas breakdown voltage; P is the gas pressure; d is the distance between the electrodes; γ is the secondary electron emission coefficient; and A and B are gas related constants.

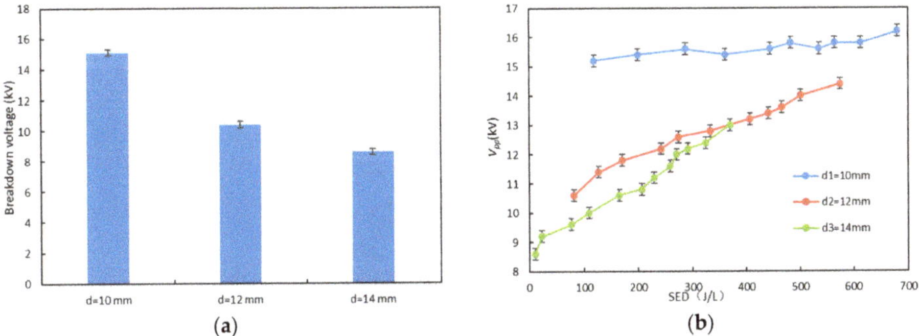

(a)
(b)

Figure 11. (**a**) The influence of electrode diameter on breakdown voltage; and (**b**) the relationship between SED and discharge voltage for different electrode diameters.

It can be seen from Equation (9) that increasing the diameter of the inner electrode can reduce the distance between the electrodes (d), thus the breakdown voltage (U_b) is reduced, as shown in Figure 11a. Increasing inner electrode diameter reduces the gas gap of DBD reactor and makes gas discharge easier. Therefore, the DBD reactor with smaller gas gap has better discharge performance and higher oxidation efficiency under the same SED [38]. In addition, according to the theory of Townsend discharge, the discharge current increases exponentially with the increase of gas gap [39]:

$$I = I_0 e^{ax} \tag{10}$$

where I is the discharge current; I_0 is the initial electron flow from the cathode; x is the gas gap; and a is the first Townsend ionization coefficient.

It can be seen from Equation (10) that the larger the gas gap (x) has stronger electron avalanche effect and higher current growth rate [39]. Therefore, the DBD reactor with smaller the electrode diameter has a smaller discharge voltage change during the increase of SED, as shown in Figure 11b. This indicates that the DBD reactor with smaller electrode diameter needs to consume more SED to increase the discharge voltage and E/N. In addition, the stronger electron avalanche effect causes a greater heat loss of DBD reactor, which also increases the energy consumption. As a result, the energy efficiency of DBD reactor decreases with the decrease of electrode diameter.

However, it cannot accurately explain the experimental results only from the perspective of E/N and electron avalanche. The increase of E/N increases the electron energy, resulting in more O radicals, but also more N radicals. N radicals generate more NO via reactions R11, R14 and R15, which causes the increase of total NO_X concentration under oxygen-enriched condition. Comparing the change of NO_X concentration of DBD reactors with different electrode diameter during the discharge process (as shown in Figure 12), we can find that the NO_X concentration at the outlet of DBD reactor with 14-mm electrode diameter is lower than that at the inlet during the whole discharge process, which indicates that increasing the electrode diameter can improve the production of O radicals and inhibit the generation of NO_X at the same time.

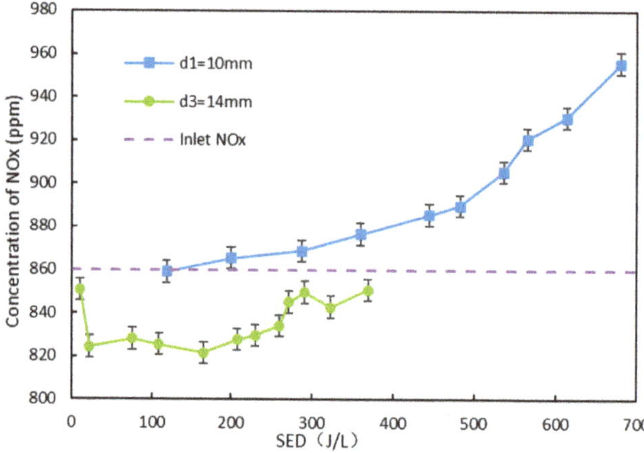

Figure 12. The change of NO_X concentration at the outlet of DBD reactor during the discharge.

To analyze the mechanism, the distribution of E/N in the gas gap of DBD reactor with different inner electrode diameter was simulated, as shown in Figure 13. The simulation results show that E/N changes (ΔE/N) in the gas gap of DBD reactors with 14-, 12- and 10-mm electrode diameter are 21.6, 51.8 and 68.5 Td, respectively. Increasing the electrode diameter does not simply improve the E/N, but also makes the distribution of E/N more centralized, which makes it easier for DBD reactor to operate at high O radical generation efficiency. It is difficult for the DBD reactor with 10-mm electrode diameter to inhibit the generation of $N_2(X, v \geq 13)$ and N radicals simultaneously due to the large ΔE/N in the gas gap. Therefore, the DBD reactor with 10-mm electrode diameter is easier to generation NO_X and has a lower NO oxidation efficiency.

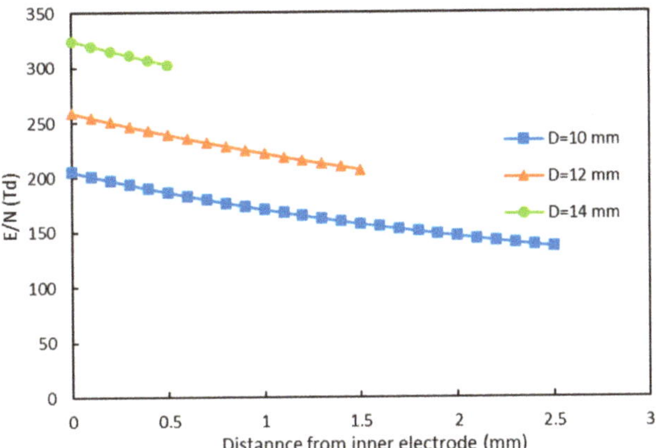

Figure 13. Distribution of E/N in gas gap of DBD reactor with different inner electrode diameter (the E/N of DBD reactor with different inner electrode diameter were calculated under their maximum oxidation degree of NO$_X$ conditions).

3.3. Effect of Inner Electrode Material

Figure 14 shows the effect of electrode material on oxidation degree of NO$_X$ in the DBD reactor with 10-mm electrode diameter (gas gap of 2.5 mm). The results show that stainless-steel electrode has the best performance on NO oxidation and the highest energy efficiency. However, this experimental result is different from results of Wang and Talebizadeh, which indicate that aluminum and tungsten electrodes have higher NO removal and energy efficiency than copper and stainless-steel electrodes due to the higher secondary electron emission coefficients γ [16,19]. γ is not a constant, which is related to the electron energy [40]. As a result, the performance of electrode material for NO treatment may also different under different E/N conditions.

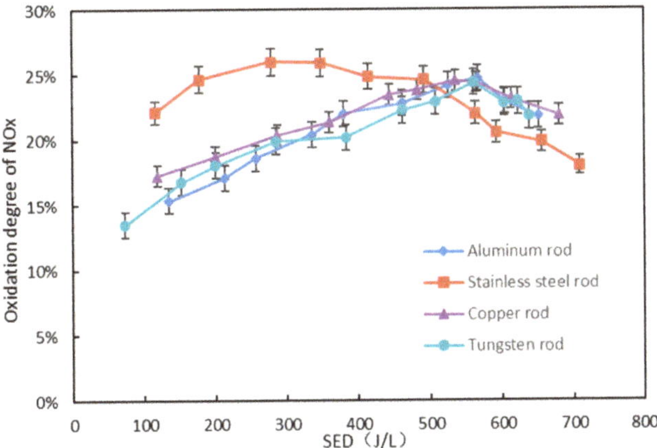

Figure 14. Effect of electrode materials on oxidation degree of NOX (electrode length is 100 mm and inner electrode diameter is 10 mm).

Therefore, we experimentally studied the influence of electrode materials on NO oxidation in the DBD reactor with 14-mm electrode diameter (gas gap of 0.5 mm), as shown in Figure 15. The experimental results show that copper electrode has a higher NO oxidation and energy efficiency than stainless-steel electrode. Obviously, the performance of electrode material for NO oxidation is different under different gas gap condition. There is obvious interaction between gas gap and electrode material. The E/N of the gas gap increases with the decrease of gas gap, which causes the increase of the mean electron energy [16]. Therefore, γ of electrode material will change under different gas gaps, resulting in the interaction between gas gap and electrode material.

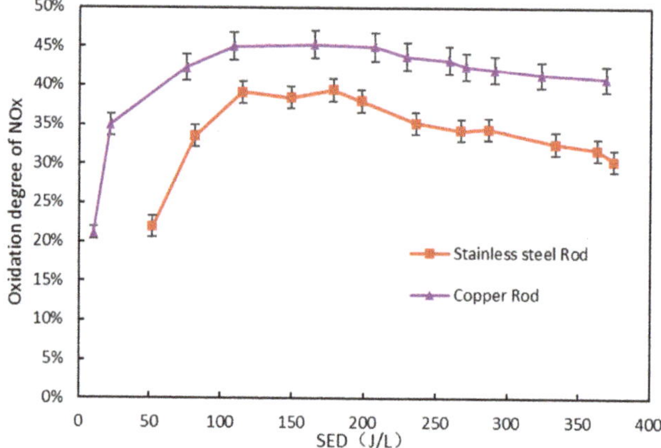

Figure 15. Effect of electrode materials on oxidation degree of NO_X (electrode length is 100 mm and inner electrode diameter is 14 mm).

3.4. Effect of Inner Electrode Shape

The effect of electrode shape on NO oxidation is shown in Figure 16. The results show that both the rod electrode and the screw electrode can make the oxidation degree of NO_X reach 45%. However, the energy efficiency of the rod electrode is higher. The experimental result is different from the result of Wang and Talebizadeh under NO/N_2 condition, as they showed screw electrode has a higher NO removal and energy efficiency [16,18,19].

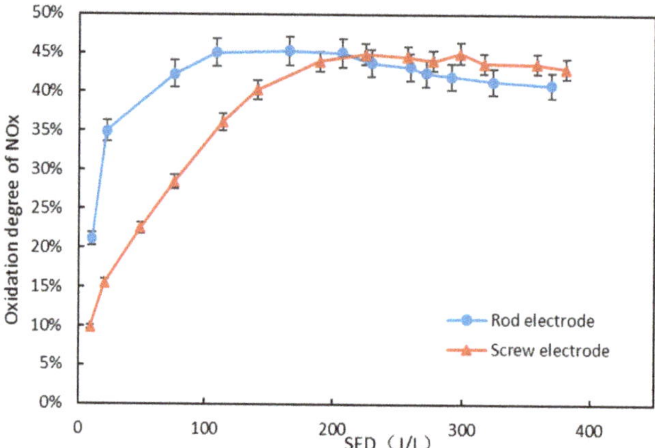

Figure 16. The effect of electrode shape on oxidation degree of NO_X under oxygen-enriched condition (electrode length is 100 mm, diameter of round rod electrode is 14 mm, diameter of external thread of threaded electrode is 14 mm, pitch is 1.5 mm and thread depth is 1 mm).

To explore the reasons, the NO, NO_2 and NO_X concentration were measured during the discharge process, as shown in Figure 17. The experiment results show that screw electrode has a higher NO and

NO_X concentration and a lower NO_2 concentration under the same SED. The distribution of E in the gas gap of DBD reactor with different inner electrode shape was simulated to analyze the mechanism, as shown in Figure 18. The simulation result shows that electric field strength (E) of the screw electrode near the top of the screw is much higher than the surface of rod electrode, which promotes the generation of N radical and the reduction efficiency of NO under NO/N_2 component [41]. However, the reduction of NO with N radicals is almost completely counterbalanced by the production of NO at 10% O_2 concentration [30,42]. Moreover, the high E/N makes the generation efficiency of O radicals decrease, as shown in Figure 9b. As a result, the NO oxidation and energy efficiency of DBD reactor with screw electrode decreases under oxygen-enriched condition.

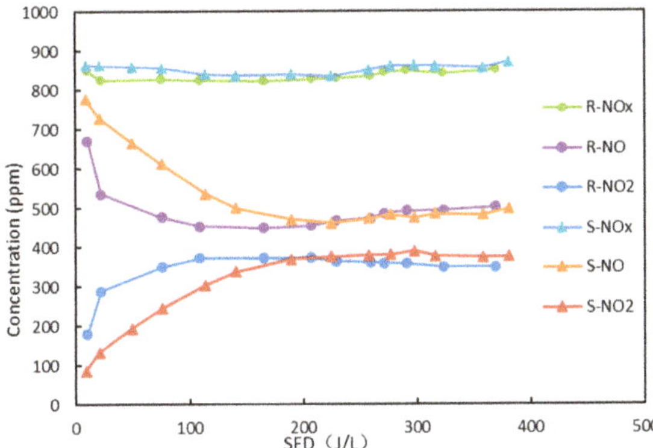

Figure 17. Effect of electrode shape on NO, NO_2 and NO_X concentration (R means, rod electrode; S, screw electrode).

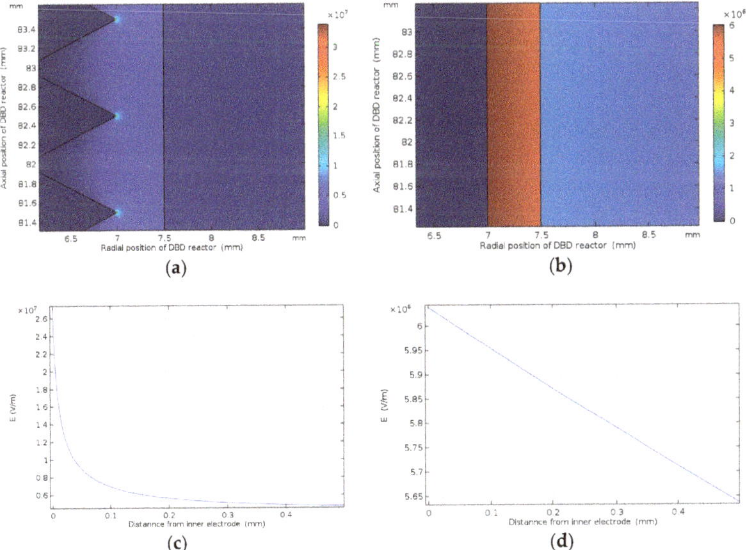

Figure 18. (a) Distribution of E in DBD reactor with screw electrode; (b) distribution of E in DBD reactor with rod electrode; (c) distribution of E in gas gap of DBD reactor with screw electrode; and (d) distribution of E in gas gap of DBD reactor with rod electrode. (Supply voltage is 5 kV, relative dielectric constant of the dielectric barrier is 3.7 and relative dielectric constant of the gas is 1.).

4. Conclusions

This paper investigates the effect of DBD structure parameters on NO oxidation, including the effects of electrode length, electrode diameter, electrode material and electrode shape on discharge characteristics and NO oxidation. The experimental results show that there is an optimal electrode length for a certain flow of gas, which makes DBD reactor have the best performance on NO oxidation and energy efficiency. Too long electrode length cannot achieve higher NO_X oxidation degree, but increases the energy consumption of DBD reactor. The oxidation degree of NO_X increases first and then decreases with the increase of SED under oxygen-enriched condition, which is different from the experimental results that NO_X removal efficiency increases with the increase of SED under the condition of no oxygen. Increasing the diameter of inner electrode not only improves the E/N, but also makes the distribution of E/N more concentrated in the gas gap, which promotes the oxidation and energy efficiency of DBD reactor. The performance of electrode material for NO oxidation is different under different gas gap condition, due to the change of E/N and secondary electron emission coefficient (γ). Compared with the rod electrode, the screw electrode has a higher electric field strength near the top of the screw, which promotes the generation of N radicals and inhibits the generation of O radicals. As a result, rod electrode has a higher NO oxidation and energy efficiency than screw electrode under oxygen-enriched condition.

Author Contributions: Conceptualization, Y.C. and L.L.; methodology, Y.C. and L.L.; software, Y.C.; validation, Y.C. and P.L.; investigation, Y.C.; resources, L.L.; data curation, Y.C.; writing—original draft preparation, Y.C.; writing—review and editing, Y.C., L.L. and P.L.; visualization, Y.C. and P.L.; supervision, L.L.; project administration, L.L.; and funding acquisition, L.L. All authors have read and agreed to the published version of the manuscript.

Funding: This research was funded by National Natural Science Foundation of China (NSFC), grant number 51679176.

Acknowledgments: The authors are thankful to all the personnel who either provided technical support or helped with data collection. We also acknowledge all the reviewers for their useful comments and suggestions.

Conflicts of Interest: The authors declare no conflict of interest.

References

1. Rahim, M.M.; Islam, T.; Kuruppu, S. Regulating global shipping corporations' accountability for reducing greenhouse gas emissions in the seas. *Mar. Policy* **2016**, *69*, 159–170. [CrossRef]
2. Boone, L. Reducing air pollution from marine vessels to mitigate arctic warming: Is it time to target black carbon? *Carbon Clim. Law Rev.* **2012**, *6*, 13–20. [CrossRef]
3. Francesco, D.N.; Claudia, C. Particulate matter in marine diesel engines exhausts: Emissions and control strategies. *Transp. Res. D: Transp. Environ.* **2015**, *40*, 166–191.
4. Fang, P.; Chen, X.B.; Tang, Z.J.; Huang, J.H.; Zeng, W.H.; Wu, H.W.; Tang, Z.X.; Cen, C.P. Current research status on air pollutant emission characteristics and control technology of marine diesel engine. *Chem. Ind. Eng. Prog (China)* **2017**, *36*, 1067–1076.
5. IMO. Nitrogen Oxides (NOx)—Regulation 13. 2019. Available online: https://www.imo.org/en/ourwork/environment/pollutionprevention/airpollution/pages/nitrogen-oxides-(nox)-%E2%80%93-regulation-13.aspx (accessed on 21 February 2019).
6. MO. Sulphur Oxides (SOx) and Particulate Matter (PM)—Regulation 14. 2019. Available online: http://www.imo.org/en/OurWork/Environment/PollutionPrevention/AirPollution/Pages/Sulphur-oxides-(SOx)-%E2%80%93-Regulation-14.aspx (accessed on 22 February 2019).
7. Fang, P.; Cen, C.; Tang, Z.; Zhong, P.; Chen, D.; Chen, Z. Simultaneous removal of SO2 and NOX by wet scrubbing using urea solution. *Chem. Eng. J.* **2011**, *168*, 52–59. [CrossRef]
8. Thagard, S.M.; Kinoshita, Y.; Ikeda, H.; Takashima, K.; Katsura, S.; Mizuno, A. Reduction for removal using wet-type plasma reactor. *IEEE Trans. Ind. Appl.* **2010**, *46*, 2165–2171. [CrossRef]
9. Lakshmipathiraj, P.; Chen, J.; Doi, M.; Takasu, N.; Kato, S.; Yamasaki, A.; Kojima, T. Electron beam treatment of gas stream containing high concentration of NOx: An in situ FTIR study. *Chem. Eng. J.* **2013**, *229*, 344–350. [CrossRef]
10. Wang, M.; Sun, Y.; Zhu, T. Removal of NOx, SO2, and Hg from simulated flue gas by plasma-absorption hybrid system. *IEEE Trans. Plasma Sci.* **2013**, *41*, 312–318. [CrossRef]

11. Ma, S.; Zhao, Y.; Yang, J.; Zhang, S.; Zhang, J.; Zheng, C. Research progress of pollutants removal from coal-fired flue gas using non-thermal plasma. *Renew. Sustain. Energy Rev.* **2017**, *67*, 791–810. [CrossRef]

12. Zhao, Z.; Yang, D.; Wang, W.; Yuan, H.; Zhang, L.; Wang, S. Electrical characters and optical emission spectra of VBD coupled SBD excited by sine AC voltage in atmospheric air. *Plasma Sci. Technol.* **2017**, *19*, 064007. [CrossRef]

13. Shekargoftar, M.; Homola, T. A new approach to the crystallization of Perovskite films by cold hydrogen atmospheric pressure plasma. *Plasma Chem. Plasma Process.* **2020**, *40*, 539–548. [CrossRef]

14. Wang, X.; Yang, Q.; Yao, C.; Zhang, X.; Sun, C. Dielectric barrier discharge characteristics of multineedle-to-cylinder configuration. *Energies* **2011**, *4*, 2133–2150. [CrossRef]

15. Wang, X.; Yao, C.; Sun, C.; Yang, Q.; Zhang, X. Numerical modelling of mutual effect among nearby needles in a multi-needle configuration of an atmospheric air dielectric barrier discharge. *Energies* **2012**, *5*, 1433–1454. [CrossRef]

16. Wang, T.; Xiao, H.P.; Zeng, J.; Duan, E.; Li, C. Effect of reactor structure in DBD for nonthermal plasma processing of NO in N2 at ambient temperature. *Plasma Chem. Plasma Process.* **2012**, *32*, 1189–1201. [CrossRef]

17. Wang, T.; Sun, B.; Xiao, H. Kinetic analysis of dielectric layer thickness on nitric oxide removal by dielectric barrier discharge. *Jpn. J. Appl. Phys.* **2013**, *52*, 46201. [CrossRef]

18. Anaghizi, S.J.; Talebizadeh, P.; Rahimzadeh, H.; Ghomi, H. The configuration effects of electrode on the performance of dielectric barrier discharge reactor for NOx removal. *IEEE Trans. Plasma Sci.* **2015**, *43*, 1944–1953. [CrossRef]

19. Talebizadeh, P.; Rahimzadeh, H.; Anaghizi, S.J.; Ghomi, H.; Babaie, M.; Brown, R. Experimental study on the optimization of dielectric barrier discharge reactor for NOx treatment. *IEEE Trans. Dielectr. Electr. Insul.* **2016**, *23*, 3283–3293. [CrossRef]

20. Li, R.; Liu, X. Main fundamental gas reactions in denitrification and desulfurization from flue gas by non-thermal plasmas. *Chem. Eng. Sci.* **2000**, *55*, 2491–2506. [CrossRef]

21. Talebizadeh, P.; Rahimzadeh, H.; Babaie, M.; Anaghizi, S.J.; Ghomi, H.; Ahmadi, G.; Brown, R. Evaluation of residence time on nitrogen oxides removal in non-thermal plasma reactor. *PLoS ONE* **2015**, *10*, 0140897. [CrossRef]

22. Kim, H.H. Nonthermal plasma processing for air-pollution control: A historical review, current issues, and future prospects. *Plasma Process. Polym.* **2004**, *1*, 91–110. [CrossRef]

23. Wang, L.; Liu, Z.; Zhu, A.; Zhao, G.; Xu, Y. Numerical Simulation of·OH and HO2· radicals in dielectric barrier discharge plasmas. *Acta Physico-Chimica Sinica* **2008**, *24*, 1400–1404.

24. Biagi Database. Available online: www.lxcat.net (accessed on 7 July 2020).

25. Itikawa Database. Available online: www.lxcat.net/Itikawa (accessed on 28 June 2020).

26. TRINITI Database. Available online: www.lxcat.net/TRINITI (accessed on 26 June 2020).

27. Kossyi, I.A.; Kostinsky, A.Y.; A Matveyev, A.; Silakov, V.P. Kinetic scheme of the non-equilibrium discharge in nitrogen-oxygen mixtures. *Plasma Sources Sci. Technol.* **1992**, *1*, 207–220. [CrossRef]

28. NIST Chemical Kinetics Database. Available online: https://kinetics.nist.gov/kinetics/index.jsp (accessed on 26 June 2020).

29. Guerra, V.; Loureiro, J. Self-consistent electron and heavy-particle kinetics in a low-pressure-glow discharge. *Plasma Sources Sci. Technol.* **1997**, *6*, 373–385. [CrossRef]

30. Zhao, G.B.; Garikipati, S.V.B.J.; Hu, X.; Argyle, M.D.; Radosz, M. The effect of oxygen on nonthermal-plasma reactions of dilute nitrogen oxide mistures in N2. *AIChE J.* **2005**, *51*, 1813–1821. [CrossRef]

31. Hagelaar, G.J.M.; Pitchford, L.C. Solving the Boltzmann equation to obtain electron transport coefficients and rate coefficients for fluid models. *Plasma Sources Sci. Technol.* **2005**, *14*, 722–733. [CrossRef]

32. Penetrante, B.M.; Hsiao, M.C.; Merritt, B.T.; Vogtlin, G.E.; Wallman, P.H.; Neiger, M.; Wolf, O.; Hammer, T.; Bröer, S. Pulsed corona and dielectric-barrier discharge processing of NO in N2. *Appl. Phys. Lett.* **1996**, *68*, 3719–3721. [CrossRef]

33. Yoshida, K.; Goto, S.; Tagashira, H.; Winstead, C.; McKoy, B.V.; Morgan, W.L. Electron transport properties and collision cross sections in C[sub 2]F[sub 4]. *J. Appl. Phys.* **2002**, *91*, 2637–2647. [CrossRef]

34. Fridman, A. *Plasma Chemistry*, 1st ed.; Cambridge University Press: Cambridge, UK, 2008; pp. 355–362.

35. Guerra, V.; Sá, P.A.; Loureiro, J. Role played by the N2(A3Σu+) metastable in stationary N2and N2-O2discharges. *J. Phys. D: Appl. Phys.* **2001**, *34*, 1745–1755. [CrossRef]

36. Zhao, G.B.; Hu, X.; Argyle, M.D.; Radosz, M. N Atom Radicals and N2(A3Σu+) Found To Be Responsible for Nitrogen Oxides Conversion in Nonthermal Nitrogen Plasma. *Ind. Eng. Chem. Res.* **2004**, *43*, 5077–5088. [CrossRef]

37. Latham, R.V. *High Voltage Vacuum Insulation: Basic Concepts and Technological Practice*, 1st ed.; Academic Press: London, UK, 1995.

38. Nemmich, S.; Tilmatine, A.; Dey, Z.; Hammadi, N.; Nassour, K.; Messal, S. Optimal Sizing of a DBD Ozone Generator Using Response Surface Modeling. *Ozone: Sci. Eng.* **2015**, *37*, 3–8. [CrossRef]

39. Zhao, Q.; Liu, S.; Tong, H. *Plasma Technology and Its Applications*, 1st ed.; National Defense Industry Press: Beijing, China, 2009.

40. Xu, X.; Zhu, D. *Gas Discharge Physics*, 1st ed.; Fudan University Press: Shanghai, China, 1996.

41. Sun, B.; Wang, T.; Yang, B.; Zhu, X.; Wang, D.; Xiao, H. Effect of Electrode Configuration on NO Removal in a Coaxial Dielectric Barrier Discharge Reactor. *J. Chem. Eng. Jpn.* **2013**, *46*, 746–750. [CrossRef]

42. Penetrante, B.M.; Bardsley, J.N.; Hsiao, M.C. Kinetic Analysis of Non-Thermal Plasmas Used for Pollution Control. *Jpn. J. Appl. Phys.* **1997**, *36*, 5007–5017. [CrossRef]

Article

Effect of Coulomb Focusing on the Electron–Atom Bremsstrahlung Cross Section for Tungsten and Iron in Nonthermal Lorentzian Plasmas

Myoung-Jae Lee [1,2], Naoko Ashikawa [3] and Young-Dae Jung [4,*]

1   Department of Physics, Hanyang University, Seoul 04763, Korea; mjlee@hanyang.ac.kr
2   Research Institute for Natural Sciences, Hanyang University, Seoul 04763, Korea
3   National Institute for Fusion Science, Toki, Gifu 509-5292, Japan; ashikawa@lhd.nifs.ac.jp
4   Department of Applied Physics, Hanyang University, Ansan, Kyunggi-Do 15588, Korea
*   Correspondence: ydjung@hanyang.ac.kr

Received: 25 June 2020; Accepted: 13 July 2020; Published: 14 July 2020

Abstract: The Coulomb focusing effect on the electron–atom bremsstrahlung spectrum is investigated in nonthermal Lorentzian plasmas. The universal expression of the cross section of nonrelativistic electron–atom bremsstrahlung process is obtained by the solution of the Thomas-Fermi equation with the effective atomic charge. The effective Coulomb focusing for the electron–atom bremsstrahlung cross section near the threshold domain is also investigated by adopting the modified Elwert-Sommerfeld factor with the mean effective charge for the bremsstrahlung process. In addition, the bremsstrahlung emission rates are obtained by considering encounters between nonthermal electrons and atoms such as Fe and W atoms. We found that the bremsstrahlung emission rates for nonthermal electron–atoms are lower than those for thermal plasmas. Various nonthermal effects on the bremsstrahlung emission rates in Lorentzian plasmas are also discussed.

Keywords: Lorentzian plasmas; coulomb focusing; bremsstrahlung

---

1. Introduction

The bremsstrahlung processes [1–12] in thermal and nonthermal plasmas have received considerable attention in astrophysics and plasma physics since the continuum X-ray spectra due to the electron–nucleus and electron–atom bremsstrahlung processes have long been used for plasma diagnostics in various weakly coupled astrophysical plasmas. Most research papers have considered the bremsstrahlung process for the electron-ion collision rather than for the electron–atom process, because of the smaller population of neutral atoms in highly ionized plasmas. Especially the electron–nucleus bremsstrahlung cross section [13] in weakly coupled plasmas has been extensively investigated by using the Yukawa-type Debye-Hückel potential. The electron–atom bremsstrahlung process [9,14,15] is quite important in weakly ionized plasmas, i.e., where the scattering of electrons by neutral atoms is important. In the electron–atom bremsstrahlung or scattering case, the screening [1,9] by the bound atomic electrons plays a crucial role in the cross section and the spectrum of the bremsstrahlung process. The influence of atomic screening on the electron–atom bremsstrahlung cross section can be studied with the Thomas-Fermi statistical model [16] for many-electron atoms. In addition, the doubly differential electron–atom bremsstrahlung cross section with form factor based on the approximate self-consistent Dirac-Hartree-Fock-Slater calculations has been obtained at high energies [15]. It is obvious that the Coulomb focusing factor for the electron-ion bremsstrahlung process is not the same with the electron–atom bremsstrahlung process because the effective charge of the bound atomic electrons is different from the total charge of the nucleus. However, the influence of Coulomb focusing on the electron–atom bremsstrahlung cross section and the radiation spectrum has not been investigated yet. In many plasmas, the coupling of the thermal plasma with the external radiation field produces the nonthermal plasma whose distribution is quite different from that of a Maxwellian plasma [17,18]. It has also been shown that the plasma electrons that deviated from the Maxwellian distribution are well represented by the Lorentzian distribution since the interaction potential in nonthermal plasmas cannot be appropriately obtained by using the conventional Debye-Hückel potential [18–21]. Therefore, we are motivated to study the bremsstrahlung process for electron–atom collision under the influence of atomic screening and Coulomb focusing correction in nonthermal plasmas. In this work, we derive the bremsstrahlung cross section for the electron–atom system by employing the Thomas-Fermi method with the effective charge of the atom. The Coulomb focusing correction for the electron–atom bremsstrahlung cross section near the threshold domain is also obtained by the modified Elwert-Sommerfeld factor with the mean effective charge for the electron–atom interaction based on the Thomas-Fermi solution. In addition, we obtain the Coulomb focused bremsstrahlung emission rates by encounters of nonthermal electrons and neutral atoms such as Fe (iron) and W (tungsten) atoms.

This paper is composed as follows: in Section 2, we introduce the Thomas-Fermi model for atoms and discuss the solution of the Thomas-Fermi model. We also obtain the mean effective charge based on the Thomas-Fermi method. In Section 3, we discuss the nonrelativistic electron–atom bremsstrahlung process and the Elwert-Sommerfeld Coulomb focusing factor. In Section 4, we obtain the modified electron–atom bremsstrahlung cross section by using the Thomas-Fermi solution including the Coulomb focusing factor with the mean effective charge of many-electron atoms. In Section 5, we obtain the closed form of the bremsstrahlung emission rates in nonthermal plasmas using the Lorentzian distribution function. In Section 6, the influence of nonthermal plasma on the electron–atom bremsstrahlung spectrum is investigated in nonthermal plasmas represented by the Lorentzian distribution function. Finally, the conclusions are given in Section 7.

2. Mean Effective Charge

The electron–atom interaction potential $V_{e-a}(\mathbf{r})$ for the electron–atom bremsstrahlung process is represented by:

$$V_{e-a}(\mathbf{r}) = -\frac{Ze^2}{r} - \int d^3\mathbf{r}' \frac{e\rho_e(\mathbf{r}')}{|\mathbf{r} - \mathbf{r}'|}, \tag{1}$$

where $\mathbf{r}$ and $\mathbf{r}'$ denote the position of the projectile electron with respect to the center of the target atom and the position of the bound electron, respectively, Z is the charge number of the nucleus, e is the electron charge, and $\rho_b(\mathbf{r}')[=-en_b(\mathbf{r}')]$ is the bound electron density of the target atom with $n_b(\mathbf{r}')$ being the bound electron number density. Then, the Fourier transformation $\widetilde{V}_{e-a}(\mathbf{q})$ of Equation (1) is obtained as:

$$\begin{aligned}
\widetilde{V}_{e-a}(\mathbf{q}) &= \int d^3\mathbf{r}\, e^{-i\mathbf{q}\cdot\mathbf{r}}\, V_{e-a}(\mathbf{r}) \\
&= -\frac{4\pi e^2}{q^2}\left[Z - F_a(\mathbf{q})\right],
\end{aligned} \tag{2}$$

where $F_a(\mathbf{q})$ denote the atomic form factor due to the distribution of bound atomic electrons given by $F_a(\mathbf{q}) = \int d^3\mathbf{r}'\, e^{-i\mathbf{q}\cdot\mathbf{r}}n_b(\mathbf{r}')$. It is obvious that $F_a(\mathbf{q})$ is zero in the electron-ion bremsstrahlung process since there are no atomic electrons in the target ion. The Thomas-Fermi model [22] is very useful to study the process in the many-electron atoms and the collision dynamics in the neutral atoms. Using the typical parameters in the Thomas-Fermi model, $F_a(\bar{q})$ is represented by:

$$\begin{aligned}
F_a(\bar{q}) &= \frac{4\pi}{q}\frac{b^2}{Z^{2/3}}\int_0^\infty dx\, x\sin(\bar{q}x)\, n_b(x) \\
&= Z[1 - \bar{q}H(\bar{q})],
\end{aligned} \tag{3}$$

where the parameter $b = (1/2)(3\pi/4)^{2/3}a_0$, $a_0(=\hbar^2/me^2)$ is the Bohr radius of the hydrogen atom, $\hbar$ is the Planck constant divided by 2π, m is the electron mass, $\bar{q}\,(\equiv qb/Z^{1/3})$ is the dimensionless momentum transfer, $x\,(\equiv r'Z^{1/3}/b)$ is the dimensionless distance, and $n_b(x)$ is the Thomas-Fermi number density given by:

$$n_b(x) = \frac{Z^2}{4\pi b^3}\left[\frac{X(x)}{x}\right]^{3/2},\tag{4}$$

with $X(x)$ being the solution of the Thomas-Fermi equation, $d^2X/dx^2 = X^{3/2}/x^{1/2}$ (with the boundary condition $X(0) = 1$), and $H(\bar{q})$ is the screening function of the atomic electrons in the Thomas-Fermi scheme defined by:

$$H(\bar{q}) = \int_0^\infty dx \, \sin(\bar{q}x)X(x).\tag{5}$$

The approximate Thomas-Fermi solution can be often given in the form of single-exponential Mott-Massey solution [9,16], such as:

$$X_{MM}(x) \cong e^{-sx},\tag{6}$$

where $s \cong 0.66$. Since the boundary conditions [16] for the Thomas-Fermi equation, $d^2X/dx^2 = X^{3/2}/x^{1/2}$, are known as $X(0) = 1$ and $X(\infty) = 0$, the single-exponential approximate Mott-Massey solution, $X_{MM}(x) \cong e^{-sx}$, would be reasonably reliable to investigate the atomic collision and radiation processes including many-electron neutral atoms. By taking of the Mott-Massey's Thomas-Fermi solution given by Equation (6), the atomic form factor is obtained as:

$$F_a(\bar{q}) = Z\left(1 - \frac{\bar{q}^2}{\bar{q}^2 + s^2}\right).\tag{7}$$

Now, Equation (2) based on Mott-Massey's single-exponential Thomas-Fermi solution becomes:

$$\widetilde{V}'_{e-a}(\bar{q}) = -\frac{4\pi Ze^2}{\bar{q}^2}\left(\frac{b}{Z^{1/3}}\right)^2\frac{\bar{q}^2}{\bar{q}^2 + s^2}.\tag{8}$$

Hence, the effective charge of the target atom can take the form:

$$\begin{aligned}Z_{eff}(\bar{q}) &= Z - F_a(\bar{q})\\[2mm] &= Z\frac{\bar{q}^2}{\bar{q}^2 + s^2}.\end{aligned}\tag{9}$$

The mean effective charge $\overline{Z}_{eff}$ for the electron–atom bremsstrahlung process is then given by:

$$\begin{aligned}\overline{Z}_{eff}(\bar{q}_m) &= Z\frac{\bar{q}_m^2}{\bar{q}_m^2 + s^2}\\[2mm] &= \frac{Z q_m^2}{q_m^2 + s^2 Z^{2/3}/b^2},\end{aligned}\tag{10}$$

where $\bar{q}_m$ is defined by $\bar{q}_m(= q_m b/Z^{1/3}) = (1/2)(3\pi/4)^{2/3}Z^{2/3}$ and q_m is the maximum momentum transfer given by $q_m \approx 1/a_Z$, with $a_Z(= a_0/Z)$ being the Bohr radius of the hydrogenic ion and Ze being the nuclear charge. Here, it is expected that Equation (10) is the universal expression of the mean effective charge $\overline{Z}_{eff}$ for the bremsstrahlung and collision processes with many electron atoms since the effective charge is obtained by the Thomas-Fermi solution and the maximum momentum transfer is given by the main contribution region for the binary-encounter. Since the mean effective charge $\overline{Z}_{eff}$ has a universal expression, Equation (10) can be the general expression of the effective charge for collision and radiation processes including neutral atoms with nuclear charge Ze.

3. Electron–Atom Bremsstrahlung and Coulomb Focusing

Using the second-order nonrelativistic perturbation analysis [8,14], the differential electron-ion bremsstrahlung cross section $d^2\sigma_b$ can be written as:

$$d^2\sigma_b = d\sigma_C \cdot dW_\omega,\tag{11}$$

where $d\sigma_C(q)$ is the differential elastic scattering cross section:

$$d\sigma_C(q) = \frac{1}{2\pi\hbar v_0^2}\left|\widetilde{V}(\mathbf{q})\right|^2 q\,dq,\tag{12}$$

v_0 is the initial velocity of the projectile electron, $\widetilde{V}(\mathbf{q})$ is the Fourier transformation of the interaction potential $V(\mathbf{r})$:

$$\widetilde{V}(\mathbf{q}) = \int d^3r\, e^{-i\mathbf{q}\cdot\mathbf{r}}\, V(\mathbf{r}),\tag{13}$$

where $\mathbf{q}(=\mathbf{k}_0 - \mathbf{k}_f)$ is the momentum transfer, and $\mathbf{k}_0$ and $\mathbf{k}_f$ are the wave vectors of the initial and final states of the projectile electron, respectively. Here, $dW_\omega/d\Omega$ is the differential probability of emitting a photon of frequency between ω and $\omega + d\omega$ in the solid angle $d\Omega$:

$$\frac{dW_\omega}{d\Omega} = \frac{\alpha}{4\pi^2}\Lambda^2 \sum_{\hat{e}}\left|\hat{e}\cdot\mathbf{q}\right|^2 \frac{d\omega}{\omega},\tag{14}$$

where Λ is the Compton wave number given by $\Lambda = \hbar/mc$, α is the fine structure constant, and $\hat{e}$ is the unit photon polarization vector. By integrating over the directions of the radiation photon in Equation (14), we obtain the bremsstrahlung cross section in the form:

$$d^2\sigma_b(q) = \frac{1}{3\pi^2\beta_0^2}\frac{\alpha}{(mc^2)^2}\left|\widetilde{V}(\mathbf{q})\right|^2 q^3\, dq\,\frac{d\omega}{\omega},\tag{15}$$

since the summation over polarizations gives the angular distribution factor $\sin^2\theta$, where θ is the angle between $\mathbf{k}_0$ and $\mathbf{q}$. In Equation (15), the quantity β_0 is defined as $\beta_0 = v_0/c$. In the nonrelativistic Born approximation, it is known that the domain of applicability is $v_0 > Z\alpha c$ since the projectile energy $E_0(\equiv mv_0^2/2)$ is greater than $Z^2 Ry$, where $Ry(= me^4/2\hbar^2 \int \approx 13.6\,\text{eV})$ is the Rydberg constant. It has been also known that the nonrelativistic Bethe-Heitler formula is invalid for the final state of the projectile electron near the cutoff region $mv_0^2/2 \int \approx \hbar\omega$ owing to the inaccuracy of the Born approximation for $v_0 \approx Z\alpha c$. Therefore, the nonrelativistic Bethe-Heitler formula must be corrected for hard spectral photon energies. In order to correct the Bethe-Heitler cross section, we must consider the motion of the initial and final states of the projectile electron in the external field of the target ion using a continuum wave function for the Coulomb potential since the final Coulomb wave function must be different from the initial Coulomb wave function due to the momentum transfer and the energy loss of the initial projectile electron. It has been shown that the Coulomb correction to the nonrelativistic Bethe-Heitler bremsstrahlung formula [1,3] using the Born approximation is obtained by the Elwert-Sommerfeld factor which is given by the ratio of the absolute square of the Coulomb wave functions at infinity $(r \to \infty)$ and at the origin $(\mathbf{r} = 0)$. The Coulomb correction in the Hamiltonian transition matrix element can be well approximated by the ratio of the absolute square of the final Coulomb s-wave function $[22]\left|\Psi_f(0)\right|^2[= \pi\eta_f e^{\pi\eta_f}/\sinh(\pi\eta_f)]$ to the initial Coulomb s-wave wave function $\left|\Psi_i(0)\right|^2[= \pi\eta_i e^{\pi\eta_i}/\sinh(\pi\eta_i)]$ at the origin, since $l = 0$ survives for $r \to 0$ and the mutual Coulomb interaction between the electron and the target ion is quite effective for the small separation due to the Coulomb focusing effect, where $\eta_f = Ze^2/\hbar v_f$ and $\eta_i = Ze^2/\hbar v_0$. Therefore, the square of the Coulomb wave function at the origin, $r = 0$, when the incident amplitude of the wave normalized to unity at infinity, $r \to \infty$, is given by $\left|\Psi(0)\right|^2[= \pi\eta e^{\pi\eta}/\sinh(\pi\eta)]$. It is quite obvious that the main contribution to emission due to the bremsstrahlung process comes from the wave near the center of the scattering system since the acceleration of the projectile electron is largest near the scattering center. Hence, the Coulomb correction known as the Elwert-Sommerfeld Coulomb focusing factor [3,14,23] is represented by:

$$f_{CF}(\eta_i, \eta_f) = \frac{|\Psi_f(0)|^2}{|\Psi_i(0)|^2} = \frac{\eta_f}{\eta_i} \frac{1 - \exp(-2\pi\eta_i)}{1 - \exp(-2\pi\eta_f)}, \tag{16}$$

and $f_{CF}(\eta_i, \eta_f) \to 1$ in the Born limit, i.e., $\eta_i >> 1$ and $\eta_f >> 1$. It has been also shown that the Elwert-Sommerfeld factor can correctly modified the electron-impact excitation cross section near the threshold domain [23]. As we see in Equation (15), the Coulomb focusing factor diverges at the spectral cutoff. However, this divergence compensates for the vanishing of the nonrelativistic Bethe-Heitler cross section at the cutoff, correctly resulting in a finite Bethe-Heitler cross section at the cutoff spectrum. The detailed discussion of the Coulomb correction using the Elwert-Sommerfeld Coulomb focusing factor is given in a recent work of Gould [14]. In the nonrelativistic electron–atom bremsstrahlung process, the charge number Z can be replaced by the effective charge number Z_{eff} [Equation (10)], including the influence of screening by bound electrons.

4. Coulomb Focused Bremsstrahlung Cross Section

Using Equations (4) and (15) with the integration over the momentum transfer q for the domain $q_{\min}[= (k_0 - k_f)] \leq q \leq q_{\max}[= (k_0 + k_f)]$, the electron–atom bremsstrahlung cross section [9] $(d\sigma_b/d\varepsilon)_{e-a}^{MM}$ per photon energy using Mott-Massey's single-exponential Thomas-Fermi solution $X_{MM}(x)$ is given by:

$$\left(\frac{d\sigma_b}{d\varepsilon}\right)_{e-a}^{MM} = \frac{16}{3} \frac{Z^2 \alpha r_0^2 c^2}{\varepsilon v_0^2} \left\{ \frac{1}{2} \ln\left[\frac{\bar{\xi}^2 + (\sqrt{\bar{E}_0} + \sqrt{\bar{E}_0 - \bar{\varepsilon}})^2}{\bar{\xi}^2 + (\sqrt{\bar{E}_0} - \sqrt{\bar{E}_0 - \bar{\varepsilon}})^2}\right] \right.$$
$$\left. + \frac{2\bar{\xi}^2 \sqrt{\bar{E}_0} \sqrt{\bar{E}_0 - \bar{\varepsilon}}}{[\bar{\xi}^2 + (\sqrt{\bar{E}_0} + \sqrt{\bar{E}_0 - \bar{\varepsilon}})^2][\bar{\xi}^2 + (\sqrt{\bar{E}_0} - \sqrt{\bar{E}_0 - \bar{\varepsilon}})^2]} \right\}, \tag{17}$$

where $\varepsilon(= \hbar\omega)$ is the bremsstrahlung photon energy, $r_0(= e^2/mc^2)$ is the classical electron radius, $\bar{\xi} = \xi Z^{1/3}$, $\xi(= sa_0/b) = 0.7455$, $\bar{E}_0 \equiv E_0/Ry$, and $\bar{\varepsilon} \equiv \varepsilon/Ry$. For the electron-ion bremsstrahlung process [14], the expression of the bremsstrahlung cross section $(d\sigma_b/d\varepsilon)_{e-i}$ is rather simple due to the omission of the screening effects:

$$\left(\frac{d\sigma_b}{d\varepsilon}\right)_{e-i} = \frac{16}{3} \frac{Z^2 \alpha r_0^2 c^2}{\varepsilon v_0^2} \ln\left(\frac{\sqrt{\bar{E}_0} + \sqrt{\bar{E}_0 - \bar{\varepsilon}}}{\sqrt{\bar{E}_0} - \sqrt{\bar{E}_0 - \bar{\varepsilon}}}\right). \tag{18}$$

However, the Coulomb focusing effect in the electron–atom bremsstrahlung process has not been investigated in the previous work of Jung and Lee [9], thus, the electron–atom bremsstrahlung cross section has to be corrected in the low-energy region. It is expected that Equations (6) and (15) provide the Coulomb focusing for the bremsstrahlung with neutral atoms since $\bar{Z}_{eff}$ describes the effective charge of the neutral atom by the electron-encounter since the Coulomb focusing factor is needed for the modification of the bremsstrahlung cross section near the threshold region. The modified

Elwert-Sommerfeld Coulomb focusing factor $f_{CF}(\overline{E}_0, \overline{\varepsilon}, \overline{Z}_{eff})$ for the electron–atom bremsstrahlung process is then given by:

$$
\begin{aligned}
f_{CF}(\overline{E}_0, \overline{\varepsilon}, \overline{Z}_{eff}) &= \frac{v_0}{v_f}\frac{1-\exp(-2\pi\overline{Z}_{eff}\alpha c/v_0)}{1-\exp(-2\pi\overline{Z}_{eff}\alpha c/v_f)}\\[2mm]
&= \frac{\sqrt{\overline{E}_0}}{\sqrt{\overline{E}_0-\overline{\varepsilon}}}\frac{1-\exp\left[-\frac{2\pi}{\sqrt{\overline{E}_0}}\frac{z^3}{z^2+\overline{\xi}^2}\right]}{1-\exp\left[-\frac{2\pi}{\sqrt{\overline{E}_0-\overline{\varepsilon}}}\frac{z^3}{z^2+\overline{\xi}^2}\right]},
\end{aligned}
\tag{19}
$$

since the expression of the standard Elwert-Sommerfeld factor [14] is given by $f_{CF}(\eta_i, \eta_f) = (\eta_f/\eta_i)[1-\exp(-2\pi\eta_i)]/[1-\exp(-2\pi\eta_f)]$. Hence, Equation (19) is the universal formula for the Coulomb focusing factor for the bremsstrahlung process with neutral atoms. The scaled form of the Coulomb focused nonrelativistic electron–atom bremsstrahlung cross section $(d\overline{\sigma}_b/d\overline{\varepsilon})_{e-a}^{CF}[= (d\sigma_b/d\varepsilon)_{e-a}^{MM}f_{CF}(\overline{E}_0, \overline{\varepsilon}, \overline{Z}_{eff})/(\pi a_0^2/Ry)]$ per photon energy in units of $\pi a_0^2/Ry$ including the influence of bound atomic electrons and Coulomb focusing at the spectral cutoff is then represented by:

$$
\begin{aligned}
\left(\frac{d\overline{\sigma}_b}{d\overline{\varepsilon}}\right)_{e-a}^{CF} &= \frac{16}{3\pi}\frac{Z^2\alpha^3}{\overline{\varepsilon}\sqrt{\overline{E}_0}\sqrt{\overline{E}_0-\overline{\varepsilon}}}\left\{\frac{1}{2}\ln\left[\frac{\overline{\xi}^2+(\sqrt{\overline{E}_0}+\sqrt{\overline{E}_0-\overline{\varepsilon}})^2}{\overline{\xi}^2+(\sqrt{\overline{E}_0}-\sqrt{\overline{E}_0-\overline{\varepsilon}})^2}\right]\right.\\[2mm]
&\qquad\left.-\frac{2\overline{\xi}^2\sqrt{\overline{E}_0}\sqrt{\overline{E}_0-\overline{\varepsilon}}}{[\overline{\xi}^2+(\sqrt{\overline{E}_0}+\sqrt{\overline{E}_0-\overline{\varepsilon}})^2][\overline{\xi}^2+(\sqrt{\overline{E}_0}-\sqrt{\overline{E}_0-\overline{\varepsilon}})^2]}\right\}\\[2mm]
&\quad\times\frac{1-\exp\left(-\frac{2\pi}{\sqrt{\overline{E}_0}}\frac{z^3}{z^2+\overline{\xi}^2}\right)}{1-\exp\left(-\frac{2\pi}{\sqrt{\overline{E}_0-\overline{\varepsilon}}}\frac{z^3}{z^2+\overline{\xi}^2}\right)}.
\end{aligned}
\tag{20}
$$

Since the electron–atom bremsstrahlung cross section [Equation (20)] has been obtained by the Thomas-Fermi solution $X_{MM}(x)$ and the modified Elwert-Sommerfeld Coulomb focusing factor $f_{CF}(\overline{E}_0, \overline{\varepsilon}, \overline{Z}_{eff})$ with the mean effective charge $\overline{Z}_{eff}$ for the electron–atom interaction, the modified electron–atom bremsstrahlung cross section $(d\overline{\sigma}_b/d\overline{\varepsilon})_{e-a}^{CF}$ would be quite reliable for investigating the physical properties of atomic bremsstrahlung spectra over wide range of electron energy including the spectral cutoff region. The bremsstrahlung emission rates in nonthermal Lorentzian astrophysical plasmas will also be discussed in the following section.

5. Bremsstrahlung Emission Rates in Lorentzian Kappa Plasmas

In most astrophysical and space plasmas, the external disturbances in thermal plasmas would produce the high-energy tail in the distribution of plasma electrons so that the deviations from the thermal Maxwellian distribution is expected due to the interaction between the plasma and the external perturbations. A pioneering work by Hasegawa, Mima, and Duong-van [17] showed that the nonthermal distribution due to the entropy generalization mechanism in the presence of external radiation field in astrophysical plasmas obeys the Lorentzian (kappa) velocity distribution function $f_L(v)$ [17,18,24] in the form:

$$
f_L(v) = n_e\left(\frac{m}{2\pi\kappa E_\kappa}\right)^{3/2}\frac{\Gamma(\kappa+1)}{\Gamma(\kappa-1/2)}\left(1+\frac{mv^2}{2\kappa E_\kappa}\right)^{-(\kappa+1)},
\tag{21}
$$

where v and n_e are the velocity and the density of electron, $\kappa(>3/2)$ is the spectral index in the Lorentzian distribution, $E_\kappa[\equiv(\kappa-3/2)E_M/\kappa]$ is the effective energy of the Lorentzian electrons, $E_M\equiv k_BT$, k_B is the Boltzmann constant, T is the electron temperature, and $\Gamma(z)$ represents the gamma function with the argument z. Then, the differential Lorentzian electron distribution function can be represented by $dn_L(v) = 4\pi v^2 f_L(v)dv$. It has been also shown that the radiation interaction modifies the conventional diffusion process in astrophysical plasmas so that the correction on the total diffusion coefficient can be represented by the factor $(1 + \alpha_R v^2)$, where α_R is a constant related to the external radiation field intensity, since the non-Coulombic diffusion coefficient is found to be proportional to the square of the electron velocity v and can be also induced by the interaction with the field [17]. It is also interesting to note that the Lorentzian distribution acquiesces a simple power-law form at high energies, i.e., we have $f_L(v) \propto (mv^2/2\kappa E_\kappa)^{-(\kappa+1)}$ when $mv^2/2 \gg \kappa E_\kappa$. Moreover, the nonthermal astrophysical Lorentzian distribution with the infinity spectral index κ, equivalent to the absence of the external interaction, turns out to be the thermal distribution for all velocities such as $f_L(\kappa \to \infty) \propto \exp(-mv^2/2E_{\kappa \to \infty})$ owing to the mathematical limiting relation: $\lim_{t \to \infty}(1 + x/t)^t = e^x$, where $E_{\kappa \to \infty}$ corresponds to the thermal energy in the Maxwellian plasmas, i.e., $k_B T(= E_M)$ [17]. Hence, we have found that the Lorentzian distribution $f_L(v)$ encompasses a wide range of plasma velocity distributions from the Maxwellian distribution to the inverse power law distribution. In addition, the effective Debye length λ_κ in nonthermal Lorentzian plasmas can be represented by $\lambda_\kappa = \lambda_D \mu_\kappa$ [18] where λ_D is the conventional Debye length in Maxwellian plasmas and $\mu_\kappa[\equiv \sqrt{(\kappa - 3/2)/(\kappa - 1/2)}]$ stands for the fractional measure of the nonthermal population in astrophysical Lorentzian plasmas. Hence, the bremsstrahlung emission rate P_ε for a given differential Lorentzian electron density distribution $dn_L(v_0)$ can be written by:

$$P_\varepsilon = \frac{dE_{rad}}{dVdtd\varepsilon} = \int dn_L(v_0)\, n_a\, v_0\, \varepsilon \left(\frac{d\sigma_b}{d\varepsilon}\right)_{e-a}^{MM} f_{CF}(\overline{E}_0, \overline{\varepsilon}, \overline{Z}_{eff}), \tag{22}$$

where E_{rad} is the bremsstrahlung radiation energy and n_a is the atom density. Then, the scaled bremsstrahlung emission rate $\overline{P}_\varepsilon(= P_\varepsilon/P_0)$ in units of $P_0[= (32/3\pi^{1/2})r_0^2 c n_e n_a]$ is represented by:

$$
\begin{aligned}
\overline{P}_\varepsilon = &\ \frac{Z^2}{(\kappa \overline{E}_\kappa)^{3/2}}\frac{\Gamma(\kappa+1)}{\Gamma(\kappa-1/2)}\int_{\overline{\varepsilon}}^{\infty} d\overline{E}_0 \left(1 + \frac{\overline{E}_0^2}{2\kappa \overline{E}_\kappa}\right)^{-(\kappa+1)} \\[4pt]
&\times \frac{\sqrt{\overline{E}_0}}{\sqrt{\overline{E}_0 - \overline{\varepsilon}}}\frac{1 - \exp\left(-\frac{2\pi}{\sqrt{\overline{E}_0}}\frac{z^3}{z^2 + \overline{\xi}^2}\right)}{1 - \exp\left(-\frac{2\pi}{\sqrt{\overline{E}_0 - \overline{\varepsilon}}}\frac{z^3}{z^2 + \overline{\xi}^2}\right)} \\[4pt]
&\times \left\{\frac{1}{2}\ln\left[\frac{\overline{\xi}^2 + (\sqrt{\overline{E}_0} + \sqrt{\overline{E}_0 - \overline{\varepsilon}})^2}{\overline{\xi}^2 + (\sqrt{\overline{E}_0} - \sqrt{\overline{E}_0 - \overline{\varepsilon}})^2}\right] \right. \\[4pt]
&\left. + \frac{2\overline{\xi}^2 \sqrt{\overline{E}_0}\sqrt{\overline{E}_0 - \overline{\varepsilon}}}{[\overline{\xi}^2 + (\sqrt{\overline{E}_0} + \sqrt{\overline{E}_0 - \overline{\varepsilon}})^2][\overline{\xi}^2 + (\sqrt{\overline{E}_0} - \sqrt{\overline{E}_0 - \overline{\varepsilon}})^2]}\right\},
\end{aligned}
\tag{23}
$$

where $\overline{\varepsilon}$ in the lower bound of the integral represents the cutoff, i.e., $mv_0^2/2 \approx \hbar\omega$, $\overline{E}_\kappa \equiv E_\kappa/Ry[= \overline{T}(\kappa - 3/2)/\kappa]$, and $\overline{T} \equiv k_B T/Ry$. The nonthermal effects on the bremsstrahlung emission rate as well as the electron–atom bremsstrahlung cross sections will be discussed in the following section.

6. Nonthermal and Coulomb Focusing Effects in Lorentzian (Kappa) Plasmas

The neutral elements in a plasma can be detected by using spatially resolved plasma spectroscopy [25]. Since Fe and W atoms are important elements in astrophysical and laboratory plasmas, we shall consider the electron–atom bremsstrahlung process with those elements in nonthermal Lorentzian plasmas. In order to investigate the behavior of the electron–atom bremsstrahlung in wide spectral ranges such as the soft- and hard-photon ranges, we choose $\overline{E}_0 = 15\,(\gg 1)$, i.e., $E_0 \gg Ry$.

Figure 1 shows the electron-ion bremsstrahlung cross sections $(d\overline{\sigma}_b/d\overline{\varepsilon})_{e-i}$ and the electron–atom bremsstrahlung cross sections $(d\overline{\sigma}_b/d\overline{\varepsilon})_{e-a}$ per photon energy in units of $\pi a_0^2/Ry$ as functions of the scaled photon energy $\overline{\varepsilon}$ for W and Fe atoms. In this figure, we see that the electron–atom bremsstrahlung cross section is quite different from the electron-ion bremsstrahlung cross section due to the screening effect caused by the bound electrons in the atom. We also see that the bremsstrahlung cross section is suppressed by the influence of bound atomic electrons. Figure 2 represents the scaled form of the electron–atom bremsstrahlung cross sections $(d\bar{\sigma}_b/d\bar{\varepsilon})_{e-a}$ per photon energy in units of $\pi a_0^2/Ry$ as functions of the scaled photon energy $\bar{\varepsilon}$ for W and Fe atoms. As shown in this figure, the Coulomb focusing enhances the bremsstrahlung cross section, especially for high-energy photons, for example, about 30% at $\bar{\varepsilon} = 6$. In addition, we see that the Coulomb focusing effect on the electron–atom bremsstrahlung cross section increases with an increase of the radiation photon energy. Hence, the Coulomb focused electron–atom bremsstrahlung cross sections including the influence of Coulomb focusing would be especially accurate for high-energy photons since the Coulomb focusing effect is significant near the cutoff spectral domain. Figure 3 shows the bremsstrahlung emission rate $\bar{P}_\varepsilon$ in units of $P_0[= (32/3\pi^{1/2})r_0^2 c n_e n_a]$ as a function of $\bar{\varepsilon}$ for W atom with different values of κ when the Coulomb focusing is not considered. Figure 4 shows the Coulomb focused bremsstrahlung emission rate $\bar{P}_{\varepsilon,CF}$ in units of $P_0[= (32/3\pi^{1/2})r_0^2 c n_e n_a]$ as a function of $\bar{\varepsilon}$ for W atom with different κ including the Coulomb focusing effect. Figure 5 shows $\bar{P}_\varepsilon$ as a function of $\bar{\varepsilon}$ for Fe atom with different κ excluding the Coulomb focusing effect. Figure 6 shows $\bar{P}_{\varepsilon,CF}$ as a function of $\bar{\varepsilon}$ for Fe atom with different κ including the effect of Coulomb focusing. As shown in Figures 3–6, the bremsstrahlung emission rates for the electron–atom bremsstrahlung process in Maxwellian plasmas are always greater than those in nonthermal plasmas. In addition, the nonthermal effects on the bremsstrahlung emission rate for the soft photon case are found to be more significant than those for the hard photon case. Moreover, the influence of Coulomb focusing on the bremsstrahlung emission rate decreases with a decrease of the spectral index κ. Therefore, the Coulomb focusing effect on the bremsstrahlung emission rate is more significant in thermal plasmas and in hard spectral ranges. As shown in these figures, the nonthermal effect on $\bar{P}_{\varepsilon,CF}$ decreases with decreasing $\bar{\varepsilon}$. Hence, the classification of the nonthermal character of plasmas by using the bremsstrahlung spectrum would be quite significant in hard spectral regions. It is known that the states of free-electron in a dense plasma would be blocked by electrons occupying a quantum state by using the blocking factor [26,27]. The quantum blocking effect on the bremsstrahlung spectrum in dense quantum plasmas will be treated elsewhere.

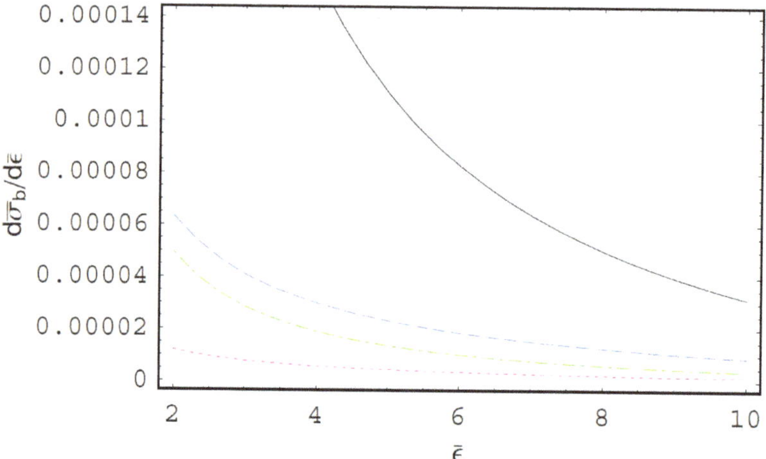

Figure 1. The scaled form of the bremsstrahlung cross section per photon energy in units of $\pi a_0^2/Ry$ as a function of the scaled photon energy $\bar{\varepsilon}$ for W and Fe atoms when $\bar{E}_0 = 15$. The black solid line is the electron-ion bremsstrahlung cross section $(d\bar{\sigma}_b/d\bar{\varepsilon})_{e-i}$ for the W atom. The blue dashed line is the electron–atom bremsstrahlung cross section $(d\bar{\sigma}_b/d\bar{\varepsilon})_{e-a}^{CF}$ for the W atom including the influence of Coulomb focusing. The green dot-dashed line is the electron-ion bremsstrahlung cross section $(d\bar{\sigma}_b/d\bar{\varepsilon})_{e-i}$ for the Fe atom. The red dotted line is the electron–atom bremsstrahlung cross section $(d\bar{\sigma}_b/d\bar{\varepsilon})_{e-a}^{CF}$ for the Fe atom including the influence of Coulomb focusing.

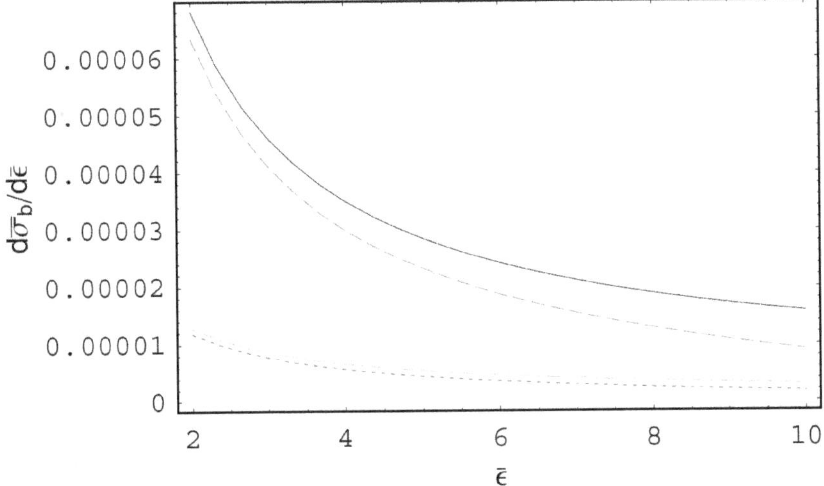

Figure 2. The scaled form of the electron–atom bremsstrahlung cross section $(d\overline{\sigma}_b/d\overline{\varepsilon})_{e-a}$ per photon energy in units of $\pi a_0^2/Ry$ as a function of the scaled photon energy $\overline{\varepsilon}$ for W and Fe atoms when $\overline{E}_0 = 15$. The black solid line is the Coulomb focused electron–atom bremsstrahlung cross section $(d\overline{\sigma}_b/d\overline{\varepsilon})_{e-a}^{CF}$ for the W atom including the influence of Coulomb focusing. The blue dashed line is the electron–atom bremsstrahlung cross section $(d\overline{\sigma}_b/d\overline{\varepsilon})_{e-a}$ for the W atom without the Coulomb focusing effect. The green dot-dashed line is the Coulomb focused electron–atom bremsstrahlung cross section $(d\overline{\sigma}_b/d\overline{\varepsilon})_{e-a}^{CF}$ for the Fe atom including the influence of Coulomb focusing. The red dotted line is the electron–atom bremsstrahlung cross section $(d\overline{\sigma}_b/d\overline{\varepsilon})_{e-a}$ for the Fe atom without the Coulomb focusing effect.

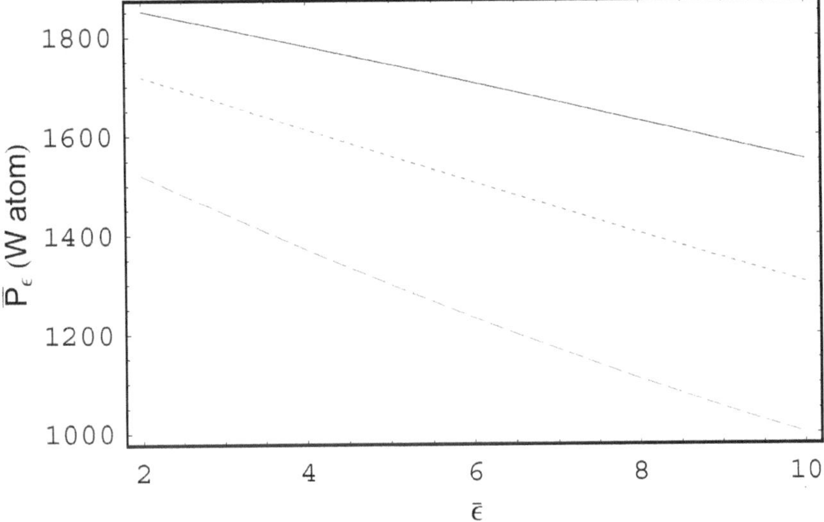

Figure 3. The bremsstrahlung emission rate $\overline{P}_\varepsilon$ in units of $P_0[= (32/3\pi^{1/2})r_0^2 c n_e n_a]$ as a function of the scaled photon energy $\overline{\varepsilon}$ for the W atom when $\overline{E}_0 = 15$ without the Coulomb focusing effect. The solid line is the case of the thermal case, i.e., $\kappa \rightarrow \infty$. The dotted line is the case of $\kappa = 3$. The dashed is the case of $\kappa = 2$.

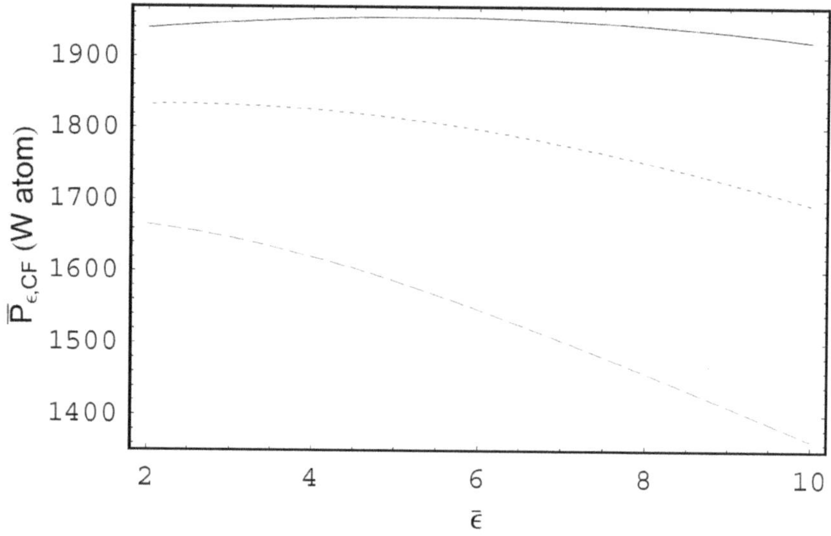

Figure 4. The Coulomb focused bremsstrahlung emission rate $\overline{P}_\varepsilon$ in units of $P_0[= (32/3\pi^{1/2})r_0^2 cn_e n_a]$ as a function of the scaled photon energy $\overline{\varepsilon}$ for the W atom when $\overline{E}_0 = 15$ including the influence of Coulomb focusing. The solid line is the case of the thermal case, i.e., $\kappa \to \infty$. The dotted line is the case of $\kappa = 3$. The dashed is the case of $\kappa = 2$.

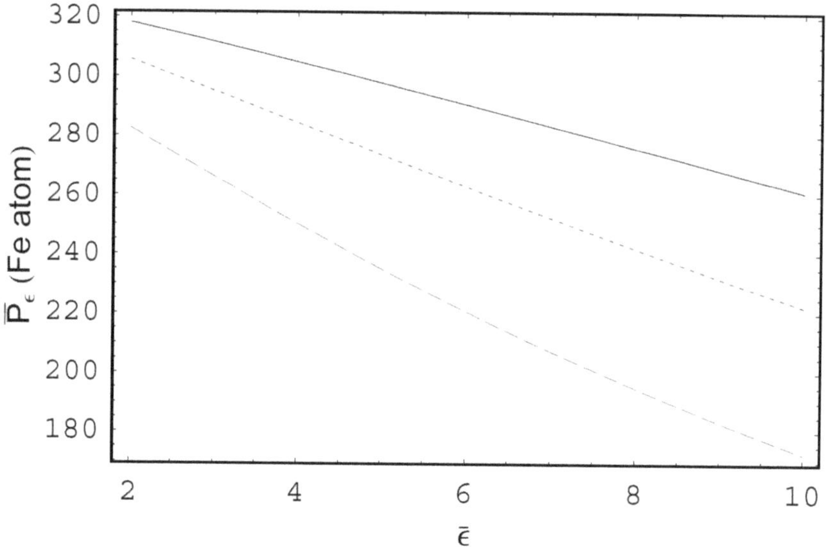

Figure 5. The bremsstrahlung emission rate $\overline{P}_\varepsilon$ in units of $P_0[= (32/3\pi^{1/2})r_0^2 cn_e n_a]$ as a function of the scaled photon energy $\overline{\varepsilon}$ for the Fe atom when $\overline{E}_0 = 15$ without the Coulomb focusing effect. The solid line is the case of the thermal case, i.e., $\kappa \to \infty$. The dotted line is the case of $\kappa = 3$. The dashed is the case of $\kappa = 2$.

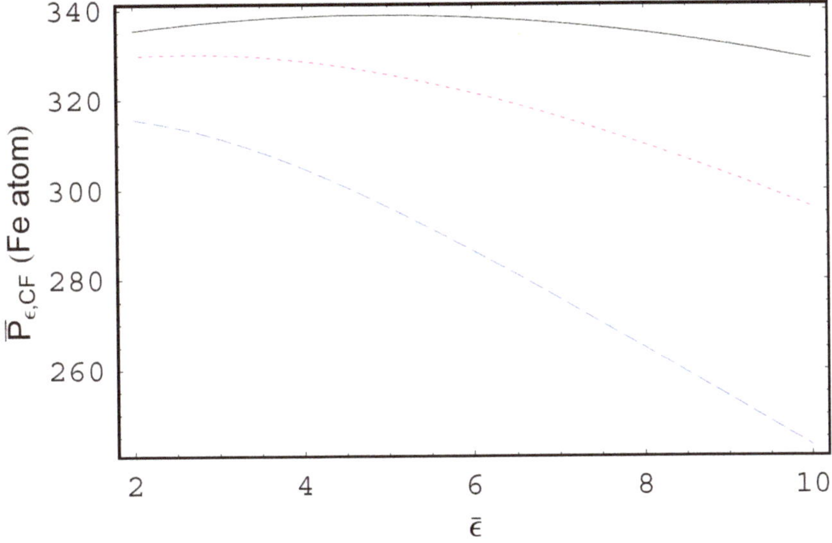

Figure 6. The Coulomb focused bremsstrahlung emission rate $\overline{P}_{\varepsilon,CF}$ in units of $P_0[=(32/3\pi^{1/2})r_0^2 c n_e n_a]$ as a function of the scaled photon energy $\overline{\varepsilon}$ for the W atom when $\overline{E}_0 = 15$ including the influence of Coulomb focusing. The solid line is the case of the thermal case, i.e., $\kappa \to \infty$. The dotted line is the case of $\kappa = 3$. The dashed is the case of $\kappa = 2$.

7. Conclusions

In this work, we have investigated the Coulomb focused bremsstrahlung spectrum due to the electron–atom bremsstrahlung process in nonthermal plasmas. We derived the universal expression of the electron–atom bremsstrahlung cross section by using the Thomas-Fermi model with the effective charge method. We also derived the effective Coulomb focusing factor for the electron–atom bremsstrahlung process by using the modified Elwert-Sommerfeld factor with the mean effective charge for the binary-encounter. The Coulomb focused electron–atom bremsstrahlung cross section and the Coulomb focused bremsstrahlung emission rates in Lorentzian plasmas with Fe and W atoms were also obtained. The Coulomb focusing is found to increase the bremsstrahlung cross section. The effect of the Coulomb focusing on the electron–atom bremsstrahlung cross section becomes bigger with an increase of the radiation photon energy. Moreover, the bremsstrahlung emission rates for the electron–atom bremsstrahlung process in thermal Maxwellian plasmas are always greater than those in nonthermal Lorentzian plasmas. The nonthermal effect on the bremsstrahlung emission rate for the soft photon case is more significant than those for the hard photon case. Hence, it is expected that the hard-photon X-ray spectroscopy of the electron–atom bremsstrahlung process would be useful to explore the physical properties of nonthermal plasmas. In addition, we have found that the Coulomb focusing effect on the bremsstrahlung emission rate is important in thermal plasmas and in hard spectral ranges. It was shown that the synchrotron radiation image would be important for the investigation of the degree of anisotropy in fusion plasmas [28]. It was also shown that the screening effect plays a significant role in the photoionization process in weakly coupled plasmas [29,30]. Hence, the Coulomb focusing effects on the bremsstrahlung process in magnetized plasmas and on the photoionization process will also be investigated elsewhere by using the modified effective charge method [31,32]. Recently, the physical significance of dusty plasmas has received a considerable attention since the dusty plasmas can be found in various astrophysical complex plasmas as well as in laboratory plasma devices [33–38]. The investigation of the influence of Coulomb focusing on the bremsstrahlung process in dusty plasmas will also be treated elsewhere since the charging of the dust grains takes a crucial role in the bremsstrahlung spectrum. It is quite obvious that the simple universal theoretical model such as the Thomas-Fermi model [39] provides a necessary intellectual framework for the collision and radiation processes as well as the geometrical configurations of the physical system. In addition, the simple analytical model can be used to understand simulations and experiments [40,41], and to extract more physical information from them. Hence, our results for the analytic expressions of the Coulomb focused electron–atom bremsstrahlung cross section and the Coulomb focused bremsstrahlung emission rates in Lorentzian plasmas would provide the useful information on the astrophysical and laboratory nonthermal X-ray radiations. In this work, we have found that the nonthermal character of the plasma as well as the Coulomb focusing effect plays a very important role in the electron–atom bremsstrahlung process in Lorentzian plasma. These results should be useful for the investigation of radiation processes in nonthermal plasmas.

Author Contributions: Conceptualization, N.A.; Formal analysis, M.-J.L., Y.-D.J.; Funding acquisition, M.-J.L.; Investigation, Y.-D.J.; Methodology, Y.-D.J.; Project administration, M.-J.L.; Resources, N.A.; Validation, N.A.; Writing–original draft, M.-J.L., Y.-D.J.; Writing–review & editing, M.-J.L., N.A., Y.-D.J. All authors have read and agreed to the published version of the manuscript.

Funding: This research was funded by the National Research Foundation of Korea (NRF-2019M1A7A1A03088471).

Acknowledgments: Two of the authors (M.-J. Lee and Y.-D. Jung) gratefully acknowledge I. Murakami and D. Kato for their warm hospitality while visiting the National Institute for Fusion Science (NIFS), Japan. This research was carried out while two of the authors (M.-J. Lee and Y.-D. Jung) were visiting the NIFS as visiting professors. The work was supported by the National Research Foundation of Korea (NRF) grant funded by the Korean Government (NRF-2019M1A7A1A03088471).

Conflicts of Interest: The authors declare no conflict of interest.

References

1. Bethe, H.A.; Salpeter, E.E. *Quantum Mechanics of One-and Two-Electron Atoms*; Springer: Berlin/Heidelberg, Germany, 1957.
2. Bekefi, G. *Radiation Processes in Plasmas*; Wiley: New York, NY, USA, 1966.
3. Blumenthal, G.R.; Gould, R.J. Bremsstrahlung, synchrotron radiation, and Compton scattering of high-energy electrons traversing dilute gases. *Rev. Mod. Phys.* **1970**, *42*, 237. [CrossRef]
4. Gould, R.J. Low-Frequency Bremsstrahlung in Coulomb Scatterings of Nonrelativistic Electrons. *Am. J. Phys.* **1970**, *38*, 189. [CrossRef]
5. Gould, R.J. Thermal bremsstrahlung from high-temperature plasmas. *Astrophys. J.* **1980**, *238*, 1026. [CrossRef]
6. Gould, R.J. Low-energy electron-atom bremsstrahlung. *Astrophys. J.* **1986**, *302*, 205. [CrossRef]
7. Kawakami, R.; Mima, K.; Totsuji, H.; Yokoyama, Y. Bremsstrahlung from hot, dense, partially ionized plasmas. *Phys. Rev. A* **1988**, *38*, 3618. [CrossRef]
8. Gould, R.J. Multipole radiation in charged-particle scattering. *Astrophys. J.* **1990**, *362*, 284. [CrossRef]
9. Jung, Y.-D.; Lee, K.-S. Screening Effects on Nonrelativistic Bremsstrahlung in the Scattering of Electrons by Neutral Atoms. *Astrophys. J.* **1995**, *440*, 830. [CrossRef]
10. Jung, Y.-D.; Jeong, H.-D. Bremsstrahlung in electron-ion Coulomb scattering in strongly coupled plasma using the hyperbolic-orbit trajectory method. *Phys. Rev. E* **1996**, *54*, 1912. [CrossRef]
11. Krainov, V.P.; Reiss, H.R.; Smirnovi, B.M. *Radiative Processes in Atomic Physics*; Wiley: New York, NY, USA, 1997.
12. Riffert, H.; Klingler, M.; Ruder, H. Bremsstrahlung Emissivity of a Proton-Electron Plasma in a Strong Magnetic Field. *Phys. Rev. Lett.* **1999**, *87*, 3432. [CrossRef]
13. Jung, Y.-D.; Yang, K.-S. Classical Electron-Ion Coulomb Bremsstrahlung in Weakly Coupled Plasmas. *Astrophys. J.* **1997**, *479*, 912. [CrossRef]
14. Gould, R.J. *Electromagnetic Processes*; Princeton University Press: Princeton, NJ, USA, 2006.
15. Haug, E. Bremsstrahlung cross-section with screening and Coulomb corrections at high energies. *Rad. Phys. Chem.* **2008**, *77*, 207. [CrossRef]
16. Mott, N.F.; Massey, H.S.W. *The Theory of Atomic Collisions*, 3rd ed.; Oxford University Press: Oxford, UK, 1987; Volume II.

17. Hasegawa, A.; Mima, K.; Duong-van, M. Plasma Distribution Function in a Superthermal Radiation Field. *Phys. Rev. Lett.* **1985**, *54*, 2608. [CrossRef]

18. Hasegawa, A.; Sato, T. *Space Plasma Physics*; Stationary Processes; Springer: Berlin/Heidelberg, Germany, 1989; Volume 1.

19. Rubab, N.; Murtaza, G. Dust-charge fluctuations with non-Maxwellian distribution functions. *Phys. Scr.* **2006**, *73*, 178. [CrossRef]

20. Rubab, N.; Murtaza, G. Debye length in non-Maxwellian plasmas. *Phys. Scr.* **2006**, *74*, 145. [CrossRef]

21. Mendis, D.A.; Rosenberg, M. Cosmic dusty plasma. *Annu. Rev. Astron. Astrophys.* **1994**, *32*, 419. [CrossRef]

22. Bransden, B.H.; Joachain, C.J. *Physics of Atoms and Molecules*, 2nd ed.; Prentice Hall: Harlow, UK, 2003.

23. Jung, Y.-D. A simple correction for the Born approximation for electron impact excitation of hydrogenic ions. *Astrophys. J.* **1992**, *396*, 725. [CrossRef]

24. Summers, D.; Thorne, R.M. Landau damping in space plasmas. *Phys. Fluids B* **1991**, *3*, 2117.

25. Park, S.; Choe, W.; Moon, S.Y.; Park, J. Electron density and temperature measurement by continuum radiation emitted from weakly ionized atmospheric pressure plasmas. *Appl. Phys. Lett.* **2014**, *104*, 084103. [CrossRef]

26. Ebeling, W.; Fortov, V.E.; Filinov, V. *Quantum Statistics of Dense Gases and Nonideal Plasmas*; Springer: Cham, Switzerland, 2017.

27. Tallents, G.J. *An Introduction to the Atomic and Radiation Physics of Plasmas*; Cambridge University Press: Cambridge, UK, 2018.

28. Hoppe, M.; Embréus, O.; Paz-Soldan, C.; Moyer, R.A.; Fülöp, T. Interpretation of runaway electron synchrotron and bremsstrahlung images. *Nucl. Fusion* **2018**, *58*, 082001. [CrossRef]

29. Lin, C.Y.; Ho, Y.K. Influence of Debye plasmas on photoionization of Li-like ions: Emergence of Cooper minima. *Phys. Rev. A* **2010**, *81*, 033405. [CrossRef]

30. Lin, C.Y.; Ho, Y.K. Photoionization cross sections of hydrogen impurities in spherical quantum dots using the finite-element discrete-variable representation. *Phys. Rev. A* **2011**, *84*, 023407. [CrossRef]

31. Jung, Y.-D.; Gould, R.J. Energies and wave functions for many-electron atoms. *Phys. Rev. A* **1991**, *44*, 111. [CrossRef]

32. Jung, Y.-D. Screening effects for transition probabilities in collisions of charged particles with an atom or stripped ion. *Phys. Rev. A* **1994**, *50*, 3895. [CrossRef]

33. Kim, S.-H.; Merlino, R.L. Electron attachment to C7F14 and SF6 in a thermally ionized potassium plasma. *Phys. Rev. E* **2007**, *76*, 035401(R). [CrossRef] [PubMed]

34. Heinrich, J.; Kim, S.-H.; Merlino, R.L. Observations of a structure-forming instability in a dc-glow-discharge dusty plasma. *Phys. Rev. E* **2011**, *84*, 026403. [CrossRef] [PubMed]

35. Akbari-Moghanjoughi, M.; Shukla, P.K. Theory for large-amplitude electrostatic ion shocks in quantum plasmas. *Phys. Rev. E* **2012**, *86*, 066401. [CrossRef]

36. Shukla, P.K.; Akbari-Moghanjoughi, M. Hydrodynamic theory for ion structure and stopping power in quantum plasmas. *Phys. Rev. E* **2013**, *87*, 043106. [CrossRef]

37. Dzhumagulova, K.N.; Masheeva, R.U.; Ramazanov, T.S.; Donkó, Z. Effect of magnetic field on the velocity autocorrelation and the caging of particles in two-dimensional Yukawa liquids. *Phys. Rev. E* **2014**, *89*, 033104. [CrossRef] [PubMed]

38. Ramazanov, T.S.; Moldabekov, Z.A.; Gabdullin, M.T. Multipole expansion in plasmas: Effective interaction potentials between compound particles. *Phys. Rev. E* **2016**, *93*, 053204. [CrossRef]

39. Lee, M.-J.; Jung, Y.-D. Eikonal-Glauber Thomas-Fermi model for atomic collisions with many-electron atoms for plasma applications. *J. Plasma Phys.* **2018**, *84*, 905840313. [CrossRef]

40. Griem, H. *Principles of Plasma Spectroscopy*; Cambridge University Press: Cambridge, UK, 1997.

41. Fujimoto, T. *Plasma Spectroscopy*; Oxford University Press: Oxford, UK, 2004.

Article

Synthesis of Multiwalled Carbon Nanotubes on Stainless Steel by Atmospheric Pressure Microwave Plasma Chemical Vapor Deposition

Dashuai Li, Ling Tong * and Bo Gao

School of Automation Engineering, University of Electronic Science and Technology of China, Chengdu 611731, China; 201511070102@std.uestc.edu.cn (D.L.); gbo@uestc.edu.cn (B.G.)
* Correspondence: tongling@uestc.edu.cn; Tel.: +86-180-8006-5381

Received: 24 April 2020; Accepted: 26 June 2020; Published: 28 June 2020

Featured Application: Synthesis of carbon nanotubes on 304 stainless steel using ethanol as a carbon source at 500–800 °C by atmospheric pressure microwave plasma chemical vapor deposition.

Abstract: In this paper, we synthesize carbon nanotubes (CNTs) by using atmospheric pressure microwave plasma chemical vapor deposition (AMPCVD). In AMPCVD, a coaxial plasma generator provides 200 W 2.45 GHz microwave plasma at atmospheric pressure to decompose the precursor. A high-temperature tube furnace provides a suitable growth temperature for the deposition of CNTs. Optical fiber spectroscopy was used to measure the compositions of the argon–ethanol–hydrogen plasma. A comparative experiment of ethanol precursor decomposition, with and without plasma, was carried out to measure the role of the microwave plasma, showing that the 200 W microwave plasma can decompose 99% of ethanol precursor at any furnace temperature. CNTs were prepared on a stainless steel substrate by using the technology to decompose ethanol with the plasma power of 200 W at the temperatures of 500, 600, 700, and 800 °C; CNT growth increases with the increase in temperature. Prepared CNTs, analyzed by SEM and HRTEM, were shown to be multiwalled and tangled with each other. The measurement of XPS and Raman spectroscopy indicates that many oxygenated functional groups have attached to the surface of the CNTs.

Keywords: microwave plasma; AMPCVD; CNTs

---

1. Introduction

Since Iijima synthesized carbon nanotubes (CNTs) by the arc discharge process in 1991 [1], many CNT synthesis processes have been developed. Notably, the chemical vapor deposition (CVD) process has been one of the most successful methods to make multiwalled carbon nanotubes (MWCNTs) [2–5]. In order to improve the yield of CNTs, direct current plasma chemical vapor deposition (DC-PECVD) [6], radio-frequency plasma chemical vapor deposition (RF-PECVD) [7,8], and microwave plasma chemical vapor deposition (MPCVD) [9–12] were developed based on thermal CVD. Process temperatures for CVD production of CNTs typically lie in the range of 700 to 1200 °C in the pressure range of 10–10^5 Pa [2–5]. The typical temperature range for the synthesis of CNTs by PECVD is 520–1000 °C at the pressure range of 40–3000 Pa [6–11]. It has also been reported that PECVD has realized CNT synthesis at 340 °C [12]. Compared with thermal CVD, the CNT growth rates of DC-PECVD, RF-PECVD, and MPCVD are increased by 2–5 times at the same temperatures and pressures. They indicate that the plasma is very significant for the improvement of the growth rate of CNTs due to the plasma's higher activation at the high temperatures on the precursor, which enhances reaction rates.

So far, all kinds of PECVD processes work under low pressure ($<10^4$ Pa). However, the atmospheric pressure microwave plasma has the advantages of higher electron density, electron activity, electron temperature, and stronger molecular decomposition, which are very powerful in the synthesis of nanomaterials. By atmospheric pressure microwave plasma, free-standing CNTs were synthesized with 900–1500 W plasma power and 1000 °C [13–15]; the CNT growth rate increased more than ten times by thermal CVD. Most atmospheric pressured microwave plasma nanomaterial synthesis systems use waveguide plasma generators with large volumes and high power. Usually, it needs more than 900 W microwave power to maintain high-power atmospheric pressure waveguide microwave plasma. It has several problems in the nanomaterial synthesis process because of the system structure and the high power: (1) The atmosphere of the system is not closed, making the synthesis area impure; there is residual air in the synthetic environment. (2) The growth area is small, and the flow rate is high, limiting the process in nanomaterial synthesis. Nevertheless, there is still great potential in the synthesis of nanomaterials by atmospheric pressure microwave plasma.

In this paper, a kind of AMPCVD system that includes a small-scaled 200 W atmospheric pressure 2.45 GHz coaxial microwave plasma generator and a heating device is presented. The combination of atmospheric pressure microwave plasma and a CVD tube furnace makes the technology have a strong precursor decomposition, a pure nanomaterial synthesis atmosphere, a large control range of particles density, and accurate temperature control, which overcomes the pressure limits of PECVD and achieves accurate control of the nanomaterials synthesis process. For now, this is the only study of nanomaterial synthesis by AMPCVD.

The CNTs are synthesized by using AMPCVD at the temperature of 500–800 °C. Compared with the CVD and all kinds of PECVD, the CNT growth rate of AMPCVD significantly increases. Compared with the experiments in [13–15], the decomposition capacity of 200 W atmospheric pressure microwave plasma of the AMPCVD is high enough to decompose the precursors completely.

2. Materials and Methods

2.1. AMPCVD

Figure 1 shows the AMPCVD, as well as the optical detection for emission spectroscopy measurements and exhaust gas detection devices. The 200 W atmospheric pressure 2.45 GHz microwave plasma generator includes a solid-state microwave source, a coaxial plasma generator, and a gas control system. The coaxial microwave plasma generator consists of two copper tubes. The diameters of the central and outer tubes are 6 and 20 mm, respectively, and their length is 90 mm. Through the center tunnel, a mixture of Ar gas and ethanol vapor is pumped into the plasma. H_2 gas fills the space between the center and outer tubes for shaping the plasma. The solid-state microwave source provides a stable 200 W microwave power for the generator to produce atmospheric pressure plasma. The gas control system includes three mass flowmeters to give an axial gas and a swirling gas to sustain the plasma flame.

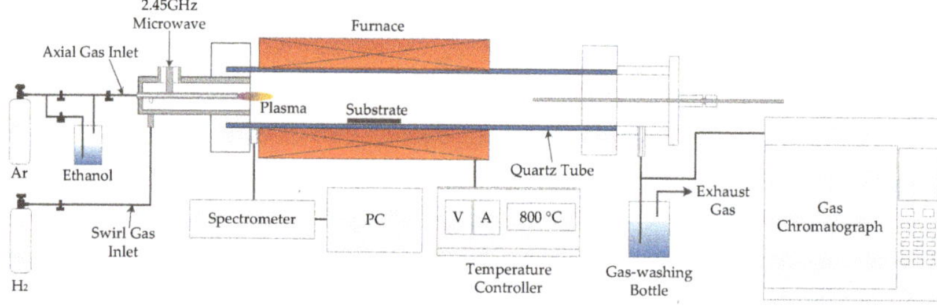

Figure 1. Schematic diagram of the atmospheric pressure microwave plasma chemical vapor deposition (AMPCVD) used for carbon nanotube (CNT) synthesis.

In the experiment, ethanol was used as the carbon source because it is cheap, easy to access, and safe—the concentration of ethanol vapor is controlled by the temperature. The 304 stainless steel sheet was used as both the substrate and the catalyst because it has a high content of iron and nickel, which are well-known catalysts for the synthesis of CNTs [16]. Briefly, 200 sccm (standard cubic centimeter per minute) argon (Ar) for the bubbling of ethanol was maintained at a temperature of 20 °C as the carbon source and 800 sccm Ar was pumped into the plasma generator through the center gas path. Via a tangential injection hole, 200 sccm hydrogen (H_2) was swirled into the plasma generator. Central two-way airflow was used to control the precursor concentration and the side hydrogen was used to stabilize the plasma and work as the deoxidizer. The exhaust gas was emitted into the air through a gas-washing bottle, and the gas-washing bottle prevented air return to the tube furnace chamber. The 304 stainless steel substrate (solid sheet, $10 \times 10 \times 0.2$ mm) was placed at the center of the high-temperature zone as the substrate for CNTs.

2.2. Experimental Setup

The pretreatment processes of the substrates are as follows: (1) Prepare a $10 \times 10 \times 2$ mm 304 stainless steel sheet as substrate. (2) Use an automatic polishing machine to polish the substrate with 80#, 800#, and 1000# sandpaper for 30 min at 120 r/min, and then polish the substrate for 1 h with cleaning cloth. (3) After polishing, sonicated the substrate in acetone, acetic acid (20%), ethanol, and deionized water for 10 min. (4) The substrate is then dried at 60 °C for 30 min and put it into the tube furnace with the polishing plate upward.

The preparatory steps are as follows: (1) Empty the air in the quartz tube (inner diameter: 52 mm, length: 600 mm) for 10 min using a vacuum pump. (2) Swirl H_2 into the apparatus with a flow of 200 sccm. When the pressure is raised to the atmosphere, the valve of the gas-washing bottle should be turn on. (3) The furnace is heated to 800 °C for 20 min to further reduce the oxygen in the quartz tube. (4) Adjust the furnace temperature to 400–1000 °C for the reaction. (5) After the furnace temperature is stable, 800 sccm Ar and 200 sccm mixture gas of Ar and ethanol vapor are pumped into the plasma generator to excite the plasma. (6) The synthesis process of CNTs lasts for 30 min, then the system is turned off. (7) Cool the apparatus down to room temperature under the H_2 environment. The CNTs are now prepared on the surface of the stainless-steel substrate.

2.3. Characterizations

In this study, an Ocean Optics MAYA-pro 2000+ optical fiber spectrometer with the spectral range of 200–1100 nm and the resolution of 1.3 nm was employed to measure the ethanol vapor microwave plasma at atmospheric pressure. Two sets of spectra were measured: the first one was for Ar-H_2 plasma, and the other one was for Ar-Ethanol-H_2. The composition from the decomposition of ethanol in the plasma emits spectral lines at a specific frequency, which are detected through analysis of the spectrum line differences between these spectral lines. The decomposition rates of the ethanol precursor at different temperatures, with and without plasma, were measured by an Agilent 6890N gas chromatograph to measure the decomposition ability of the atmospheric pressure microwave plasma. A scanning electron micrograph (SEM; Hitachi S-5000 20 kV) and a high-resolution transmission electron microscope (HRTEM; FEI Tecnai G2 F20 200 kV) were used to examine the morphology and the microstructure of the CNTs. The composition and the contents of the samples were measured by X-ray photoelectron spectroscopy (XPS; Thermo Fisher Escalab Xi+) and Raman spectroscopy (Thermo Fisher DXD).

3. Results and Discussion

3.1. Plasma Parameters

The intensity of a spectral line is proportional to the population density in the upper level of the associated transition. Variations in the strength of the lines emitted relate to the processes that take place in the plasma. Thus, the spectral was measured to determine the gas composition of the ethanol vapor microwave plasma.

Figure 2 shows the emission spectrums of Ar-H$_2$ and Ar-ethanol-H$_2$ microwave plasma. The main luminescent groups in the plasma are H$_\alpha$ (656.19 nm), H$_\beta$ (486.25 nm), H$_\gamma$ (434.09 nm), CH (389.02 nm, 431.31 nm), C$_2$ (471.06 nm, 516.08 nm, 563.1 nm), and OH (308.75 nm). The intensity of H$_\alpha$ lines in both spectrums is almost equal, indicating a high H radical concentration in the plasma. H radicals can effectively etch the sp^2 carbon phase and graphite phase, which is conducive to the preparation of high-purity CNTs. The relative intensity of the spectral lines between 690 and 900 nm are all Ar I lines and almost equal strength between both spectrums, indicating that the ethanol does not change the energy state of the microwave plasma.

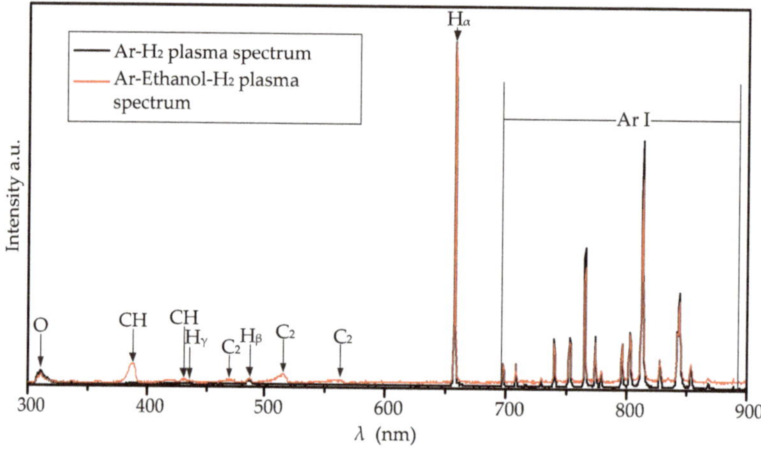

Figure 2. The measured spectrums of Ar-H$_2$ and Ar-ethanol-H$_2$ microwave plasma in atmospheric pressure.

3.2. Exhaust Gas Detection

The experiment was carried out to understand the role of plasma in the decomposition of precursors. The research was divided into two groups: one was an experimental group with the 200 W atmospheric pressure microwave plasma, and the other was without plasma. When the plasma is turned off at room temperature, we measure the initial concentration of the exhaust gas and then use gas chromatography to detect the specific density of the exhausted gas treated by the different temperatures of 400, 500, 600, 700, 800, 900, and 1000 °C, with and without plasma. The ethanol content rates are obtained by comparing the ethanol content at different temperatures with the original content. Through the comparative analysis of exhaust gas composition at different temperatures, the roles of microwave plasma and temperature can be figured out.

Figure 3 shows the ethanol content rates at the temperature of 400, 500, 600, 700, 800, 900, and 1000 °C, with and without microwave plasma. When the plasma is turned off, the apparatus will become an atmospheric pressure CVD; the ethanol content rates decreased with the increase of furnace temperatures. When the furnace temperature was as high as 1000 °C, the ethanol content rate was 26.86%. However, the ethanol content rate was all about 1% at any furnace temperature when the microwave plasma was turned on. In the experiment, no other organic hydrocarbon except ethanol was found in the gas chromatograph test results, indicating that the active group did not recombine into hydrocarbons. Therefore, we can conclude that the 200 W atmospheric pressure microwave plasma almost completely decomposes the precursor.

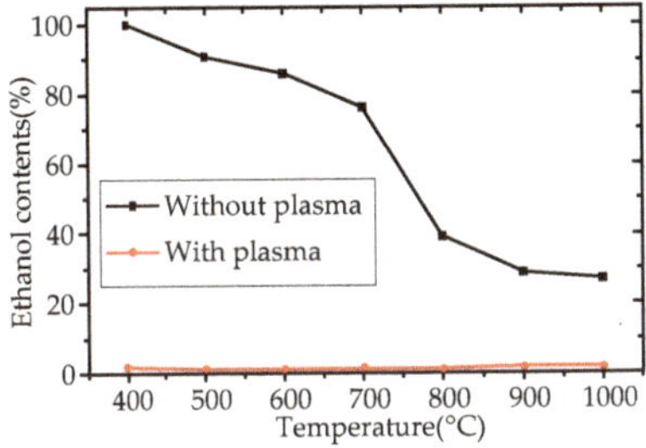

Figure 3. The ethanol contents at different temperatures in the exhaust gas, with and without plasma treatment.

3.3. CNTs

Figure 4 shows the Raman spectra of the samples grown at 500, 600, 700, and 800 °C on the surface of the substrate with the laser of 514 nm at room temperature. In all curves, two peaks centered at 1360 and 1593 cm^{-1} were observed. Both of them correspond to the D and G bands of the graphitic phase, indicating the presence of crystalline graphitic MWCNTs [16]. With the increase of the synthesis temperature, the D peak narrowed and the G peak heights increased. The intensity ratio of D peak and G peak (I_D/I_G) is sensitive to structural defects in the MWCNTs. The smaller the ratio, the higher the degree of graphitization. The I_D/I_G ratio decreased with the increase of the synthesis temperature from 600 to 800 °C; the lowest I_D/I_G ratio (1.0) was found at 800 °C. The G peak of pure graphite is at 1582 cm^{-1}; the defects in the samples caused the blue shift of G peak, verifying that the products' defect concentration is relatively high. From the view of the HRTEM, the products are mainly nanotubes, as shown in Figure 6; we can prove that the products are CNTs. The I_D/I_G ratio of CNTs grown on stainless steel substrate by PECVD [17–21] is 1.0–1.5, similar to that by AMPCVD. The I_D/I_G ratio of CNTs grown on 316 stainless steel pretreated by oxidation processes is 0.48–0.56 at 800–900 °C [16], indicating that the stainless steel substrate mainly determines the quality of the CNTs and the quality of CNTs synthesized by AMPCVD is as good as other kinds of PECVD processes.

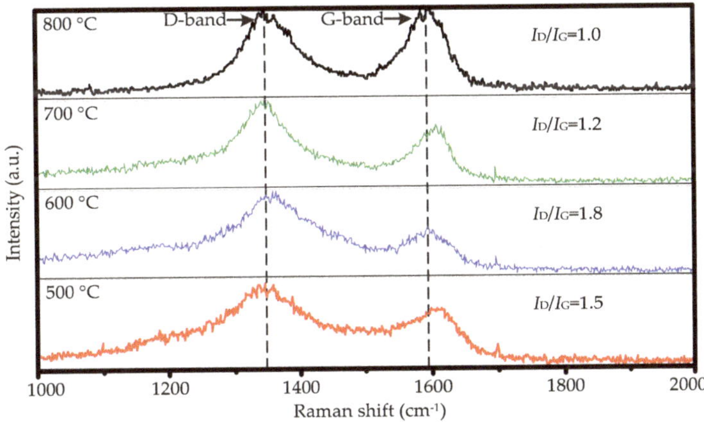

Figure 4. Raman spectra of the samples obtained at 500, 600, 700, and 800 °C on the substrate.

According to the results of plasma spectrum detection and exhaust gas detection, the ethanol precursor almost completely decomposed into active groups by 200 W atmospheric pressure microwave plasma. The recombination of carbon groups into CNTs mainly depends on temperature. We introduced the concept of conversion efficiency of carbon to describe the effect of temperature on the growth of CNTs. It represents the ratio of the carbon content of the grown CNTs to the total ethanol precursor.

Figure 5 shows the SEM micrographs and the diameter distribution of the CNTs grown on the substrate with 200 sccm ethanol precursors at the temperature of 500, 600, 700, and 800 °C. The CNTs synthesized on the stainless steel substrate lie on the substrate face, with disordered orientation and uniform distribution, consistent with the results reported in the literature [19,21], which use stainless steel substrate to grow CNTs by CVD and PECVD. It is proved that the morphology of CNTs is caused by the stainless-steel substrate. Figure 5a,d,g,j and 5b,e,h,k is the low magnification and high magnification images of the CNTs grown at the temperature of 500, 600, 700, and 800 °C, respectively. The total mass of CNTs increased with the increase of the temperature at the range of 500–800 °C. When the temperature is 800 °C, the total mass of CNTs is 300 mg. The carbon content of the total ethanol precursor is about 312 mg, indicating that approximately 96% of the precursor converted into CNTs. The precursor conversion efficiency at 500, 600, 700, and 800 °C was calculated, and they are 2%, 17%, 42%, and 96%, respectively. In the temperature range of 700–800 °C, the conversion efficiency of precursor increases significantly, indicating that the most suitable temperature for the growth of CNTs should be in the range of 700–800 °C; this needs further research.

Figure 5c,f,i,l are the diameter distribution of the CNTs grown at the temperature of 500, 600, 700, and 800 °C, respectively. With the increase in temperature, the diameter of CNTs gradually increase. The reason is that the active Fe nanoparticle, as the "seed" of the CNTs produced on the stainless-steel substrate, needs a suitable temperature range. These active nanoparticles act as catalysts during the growth of CNTs. The size and quantity of active Fe nanoparticles on the substrate increases with the increase of temperature. It makes the diameter and the precursor conversion efficiency of the CNTs increase. When the tube furnace temperature is 400 °C and 1000 °C, no CNTs grow on the substrate. The reason is that the surface activity of the stainless-steel substrate is too low to catalyze the growth of the CNTs when the temperature of the furnace is 400 °C. At 1000 °C, the surface of the stainless-steel substrate melts completely, which then cannot provide "seeds" for the growth of the CNTs.

Figure 6 shows the HRTEM micrographs of CNT ethanol dispersions dropped on a carbon-coated TEM grid. The CNTs tangle with each other, and their layers are in the range of 10–20, respectively. The wall thickness and distance between layers of the CNTs at different temperatures were measured, as shown in Figure 6. The walls of the CNTs at all temperatures are tightly aligning with a high density of graphite sheets, with a distance between layers of 0.35–0.38 nm. The layer number and wall thickness of carbon nanotubes increase with the increase of temperature. A large number of amorphous carbon layers attached to the CNTs produced at low temperatures, as shown by red arrows in Figure 6. With the increase of the temperature, the content of amorphous carbon decreased. This phenomenon is consistent with our Raman analysis results. The CNTs prepared at 700 and 800 °C consist of hollow compartments, looking like bamboo, which are not apparent at 500 and 600 °C. The carbon walls of the hollow always bulge towards the root of the CNTs. According to the vapor–liquid–solid growth method, in the growth process of CNTs, active liquid Fe nanoparticles dissolve gaseous carbon particles and then precipitate carbon atoms to form CNTs. The carbon dissolution rate of active liquid Fe nanoparticles at the temperature of 700 and 800 °C is higher than that at 500 and 600 °C, so the growth rate of CNTs is faster. When the carbon dissolution rate of Fe nanoparticles is higher than that of precipitation, new carbon layers formed, resulting in the bamboo structure.

Compared with the carbon walls of the CNTs, we found that the carbon layers of CNTs grown at high temperatures are smoother and cleaner than that grown at low temperatures. The reason is that high temperature makes the process of dissolving and precipitating carbon faster and makes it easier to grow continuous carbon layers. Figure 6j,k shows the magnified view of the joint between the wall and the hollow and the open end structure of the CNTs. XPS further characterized the CNTs prepared at 800 °C.

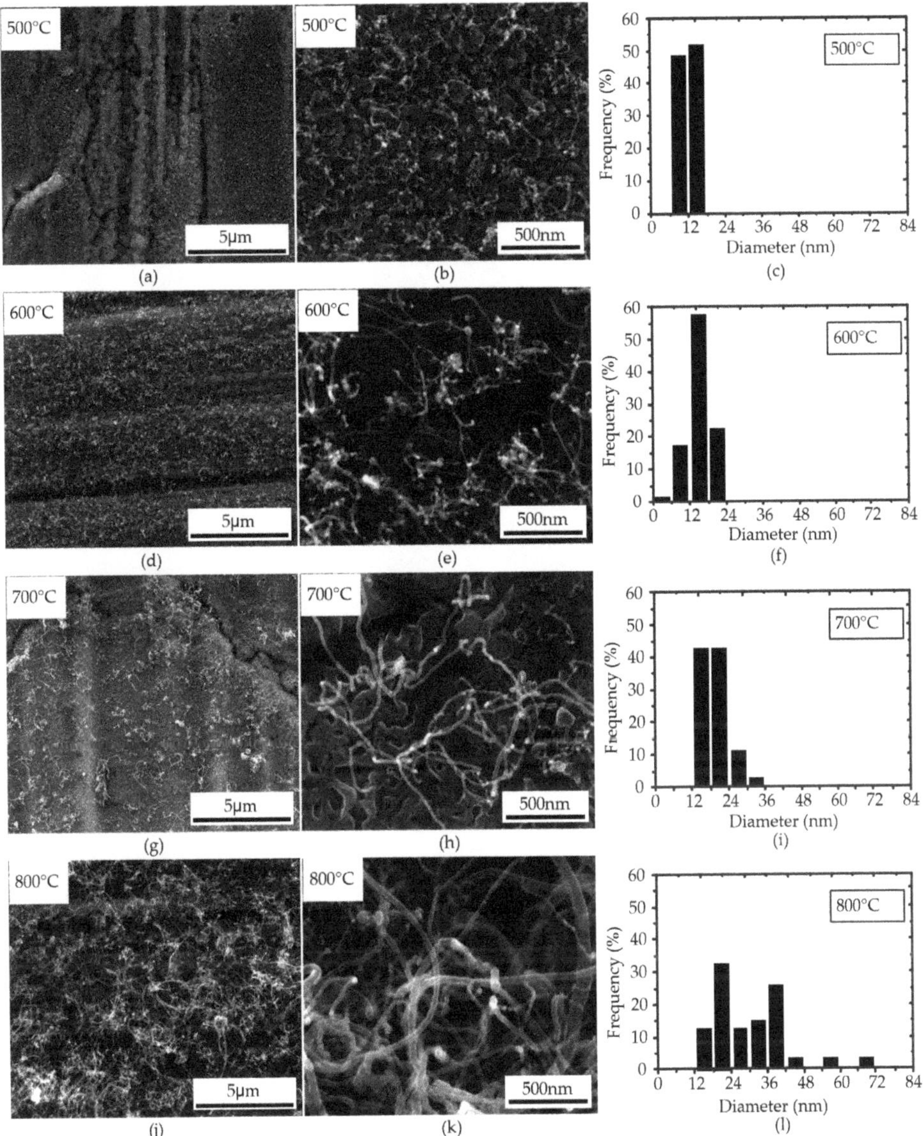

Figure 5. The SEM micrographs of the CNTs grown on the substrate with 200 sccm ethanol precursor at the temperature of 500, 600, 700, and 800 °C. (**a,d,g,j**): the low magnification images; (**b,e,h,k**): the high magnification images; (**c,f,i,l**): the diameter distribution of samples.

XPS measurement was carried out to investigate the surface functional groups of the synthesized CNTs. As shown in Figure 7a,b, the CNTs contain C, N, and O, and the magnified view of **C1s** peak indicate that they are mainly composed of C-C/C=C (284.8 eV, 59.27%), and the surface defects of CNTs are -C-OH/C-N (285.7 eV, 5.33%), -C=O (287.1 eV, 26.73%), and -COOH (288.6 eV, 8.67%). The oxygenated functional groups on the surface of CNTs were mainly from the hydroxyl groups obtained from ethanol decomposition, which caused the high defect concentration of the CNTs.

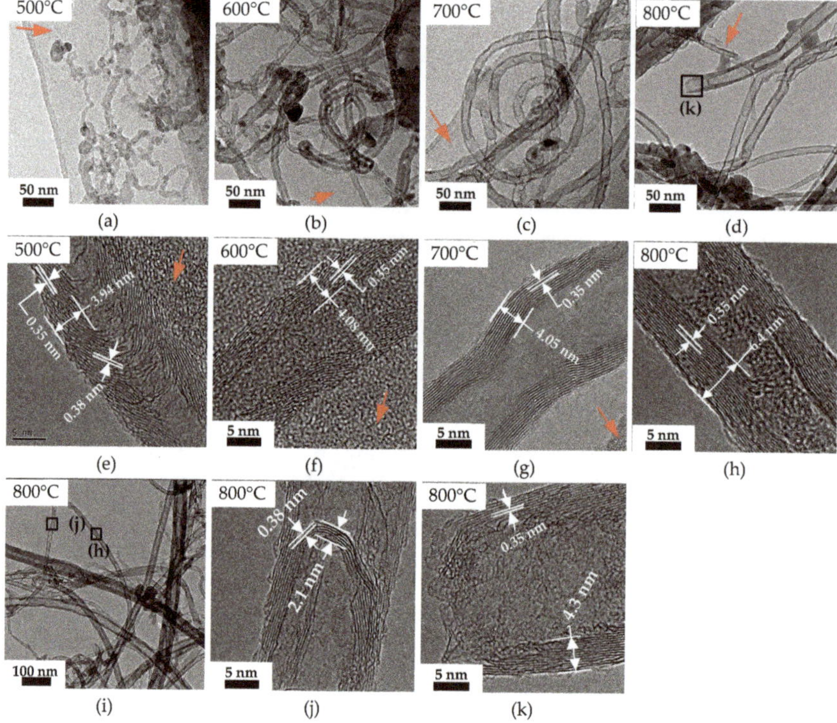

Figure 6. HRTEM images of CNTs prepared by APMCVD at the temperature of 500, 600, 700, and 800 °C. (**a–d**): low magnification images; (**e–h**): high magnification images; (**i–k**): specific structure of the CNTs grown at 800 °C.

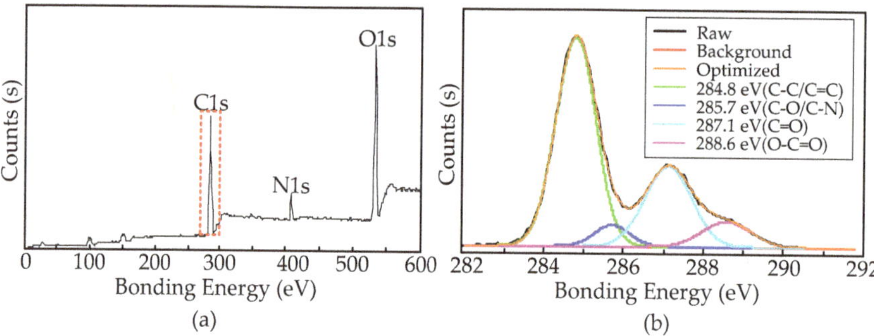

Figure 7. XPS and Raman spectra of CNTs prepared at 800 °C. (**a**): the global spectra; (**b**): the magnification view of **C1s**.

4. Conclusions

We achieved the utilization of the AMPCVD to decompose ethanol precursor by microwave plasma and the controllable synthesis of nanomaterial in a high-temperature tube furnace separately. The emission spectrum of the Ar-ethanol-H_2 microwave plasma shows that the plasma decomposed the ethanol precursor into CH, C_2, and OH groups. The results of the ethanol contents in the exhausted gas at the temperature of 400, 500, 600, 700, 800, 900, and 1000 °C, with and without plasma, also show that the decomposition capacity of 200 W atmospheric pressure microwave plasma is powerful. The CNTs were synthesized by the AMPCVD using ethanol vapor as the carbon source, with a temperature of 500, 600, 700, and 800 °C at atmospheric pressure. The tube furnace temperature controls the growth of CNTs. The Raman spectra show that the defect concentration of CNTs decreases with the rise of furnace temperature, and the quality of CNTs obtained at 800 °C is relatively high. The prepared CNTs, characterized by SEM, HRTEM, and XPS, show that the CNTs are multiwalled, tangled with each other, bamboo-shaped structured, and have a large amount of oxygen-containing functional groups on the surface, especially aldehydes. Compared with CVD, the CNTs synthesized by AMPCVD have a higher growth rate and lower defect concentration with the same substrate. Compared with DC-PECVD, RF-PECVD, and MPCVD, the quality of the CNTs synthesized by AMPCVD is similar, but with no vacuum equipment required. This indicates AMPCVD has excellent potential in nanomaterial synthesis, and further research is needed.

Author Contributions: Conceptualization and supervision, L.T., and D.L.; methodology, D.L.; validation, L.T., and B.G.; formal analysis, D.L., and L.T.; resources, D.L.; data curation, D.L.; writing—original draft preparation, D.L. and L.T.; writing—review and editing, D.L. All authors have read and agreed to the published version of the manuscript.

Funding: This work is supported by the National Key Research and Development Program of China (2016YFF0102100) and the Pre-research Project of Civil Aerospace Technology of China (D040109).

Acknowledgments: The high temperature tube furnace and sealing technology are supported by Tianjin ZHONGHUAN Electric Furnace Co. Ltd.

Conflicts of Interest: The authors declare no conflict of interest. The funders had no role in the design of the study; in the collection, analyses, or interpretation of data; in the writing of the manuscript, or in the decision to publish the results.

References

1. Iijima, S. Helical microtubules of graphitic carbon. *Nature* **1991**, *354*, 56–58. [CrossRef]
2. Ren, Z.F.; Lan, Y.C.; Wang, Y Physics, Concepts, Fabrication and Devices. In *Aligned Carbon Nanotubes*; Avouris, P., Bhushan, B., Bimberg, D., Klitzing, K., Sakaki, H., Weesendanger, R., Eds.; Springer: Berlin/Heidelberg, Germany, 2013; pp. 67–71.
3. Cheng, J.; Zou, X.P.; Yang, G.Q. Temperature Effects on Synthesis of Multi-Walled Carbon Nanotubes by Ethanol Catalyst Chemical Vapor Deposition. *Adv. Mater. Res.* **2010**, *123*, 799–802. [CrossRef]
4. Li, W.Z.; Xie, S.S.; Qian, L.X.; Chang, B.H.; Zou, B.S. Large-Scale Synthesis of Aligned Carbon Nanotubes. *Science* **1996**, *274*, 1701. [CrossRef] [PubMed]
5. Eres, G.; Puretzky, A.A.; Geohegan, D.B. In situ control of the catalyst efficiency in chemical vapor deposition of vertically aligned carbon nanotubes on predeposited metal catalyst film. *Appl. Phys. Lett.* **2004**, *84*, 1759–1761. [CrossRef]
6. Suman, N.; Mauricio, L.; Melissa, C. Synthesis and field emission properties of vertically aligned carbon nanotube arrays on copper. *Carbon* **2012**, *50*, 2641–2650.
7. Wang, Y.H.; Lin, J.; Huan, C.H. Synthesis of large area aligned carbon nanotube arrays from C2H2-H2 mixture by rf plasma-enhanced chemical vapor deposition. *Appl. Phys. Lett.* **2001**, *79*, 680–682. [CrossRef]
8. Caughman, J.B.O.; Baylor, L.R.; Guillorn, M.A.; Merkulov, V.I.; Lowndes, D.H. Growth of vertically aligned carbon nanofibers by low-pressure inductively coupled plasma-enhanced chemical vapor deposition. *Appl. Phys. Lett.* **2003**, *83*, 1207. [CrossRef]
9. Okai, M.; Muneyoshi, T.; Yaguchi, T.; Sasaki, S. Structure of carbon nanotubes grown by microwave-plasma-enhanced chemical vapor deposition. *Appl. Phys. Lett.* **2000**, *77*, 3468. [CrossRef]
10. Kuttel, O.M.; Groening, O.; Emmenegger, C.; Schlapbach, L. Electron field emission from phase pure nanotube films grown in a methane/hydrogen plasma. *Appl. Phys. Lett.* **1998**, *73*, 2113. [CrossRef]
11. Bower, C.; Zhu, W.; Jin, S.H.; Zhou, O. Plasma-induced alignment of carbon nanotubes. *Appl. Phys. Lett.* **2000**, *77*, 830. [CrossRef]
12. Xiao, Y.; Ahmed, Z.; Ma, Z.C.; Zhou, C.J.; Zhang, L.N.; Chan, M. Low Temperature Synthesis of High-Density Carbon Nanotubes on Insulating Substrate. *Nanomaterials* **2019**, *9*, 473. [CrossRef] [PubMed]
13. Chen, C.K.; Perry, W.L.; Xu, H. Plasma torch production of macroscopic carbon nanotube structures. *Carbon* **2003**, *41*, 2555–2560. [CrossRef]

14. Shin, D.H.; Hong, Y.C.; Uhm, H.S. Production of Carbon Nanotubes by Microwave Plasma-Torch at Atmospheric Pressure. *Phys. Plasmas* **2005**, *12*, 053504.

15. Lenka, Z.; Marek, E.; Ondrej, J. Characterization of Carbon Nanotubes Deposited in Microwave Torch at Atmospheric Pressure. *Plasma Process Polym.* **2007**, *4*, S245–S249.

16. Tripathi, P.V.; Durbach, S.; Coville, N.J. Synthesis of Multi-Walled Carbon Nanotubes from Plastic Waste Using a Stainless-Steel CVD Reactor as Catalyst. *Nanomaterials* **2017**, *7*, 284. [CrossRef] [PubMed]

17. Abad, M.D.; Sanchez, J.C.; Berenguer, A.; Golovko, V.B. Catalytic growth of carbon nanotubes on stainless steel: Characterization and frictional properties. *Diam. Relat. Mater.* **2008**, *17*, 1853–1857. [CrossRef]

18. Park, D.; Kim, Y.H.; Lee, J.K. Pretreatment of stainless steel substrate surface for the growth of carbon nanotubes by PECVD. *J. Mater. Sci.* **2003**, *38*, 4933–4939. [CrossRef]

19. Park, D. Synthesis of carbon nanotubes on metallic substrates by a sequential combination of PECVD and thermal CVD. *Carbon* **2003**, *41*, 1025–1029. [CrossRef]

20. Yao, B.D.; Wang, N. Carbon nanotube arrays prepared by MWCVD. *J. Phys. Chem. B* **2001**, *105*, 11395–11398. [CrossRef]

21. Hashempour, M.; Vicenzo, A.; Zhao, F.; Bestetti, M. Direct growth of MWCNTs on 316 stainless steel by chemical vapor deposition: Effect of surface nano-features on CNT growth and structure. *Carbon* **2013**, *63*, 330–347. [CrossRef]

Article

A Practical Method for Controlling the Asymmetric Mode of Atmospheric Dielectric Barrier Discharges

Ling Luo [1], Qiao Wang [1], Dong Dai [1,*], Yuhui Zhang [2] and Licheng Li [1]

[1] School of Electric Power, South China University of Technology, Guangzhou 510641, China; epluoling@mail.scut.edu.cn (L.L.); epjoking@mail.scut.edu.cn (Q.W.)

[2] Department of Electrical, Computer, and Systems Engineering, Rensselaer Polytechnic Institute, Troy, NY 12180, USA; zhangy79@rpi.edu (Y.Z.); lilc@scut.edu.cn (L.L.)

[*] Correspondence: ddai@scut.edu.cn

Received: 20 January 2020; Accepted: 11 February 2020; Published: 16 February 2020

Abstract: Atmospheric pressure dielectric barrier discharges (DBDs) have been applied in a very broad range of industries due to their outstanding advantages. However, different discharge modes can influence the stability of atmospheric DBDs, such as the density and composition of active species in discharge plasmas, thereby impacting the effect of related applications. It is necessary and valuable to investigate the control of nonlinear modes both in theoretical and practical aspects. In this paper, we propose a practical, state-controlling method to switch the discharge mode from asymmetry to symmetry through changing frequencies of the applied voltage. The simulation results show that changing frequencies can effectively alter the seed electron level at the beginning of the breakdown and then influence the subsequent discharge mode. The higher controlling frequency is recommended since it can limit the dissipative process of residual electrons and is in favor of the formation of symmetric discharge in the after-controlling section. Under our simulation conditions, the discharges with an initial driving frequency of 14 kHz can always be converted to the symmetric period-one mode when the controlling frequency is beyond 30 kHz.

Keywords: plasma; dielectric barrier discharges; state-controlling method

1. Introduction

Atmospheric dielectric barrier discharges (DBDs), due to their advantages in producing low-temperature plasmas without a precise vacuum chamber, have been widely used in a lot of industrial applications such as surface modification, energy transformation, biomedical sterilization, and pollution abatement [1–5]. Such pervasive applications have already made DBDs a prospective and an active research topic for many years.

Being a nonequilibrium dissipative system, DBDs can exhibit an abundance of nonlinear characteristics or phenomena which are extremely sensitive to the operational environment and controlled parameters [6]. Typically, it is well known that in most cases atmospheric DBDs appear as a symmetric period-one discharge (henceforth called SP1), that is, the shape and amplitude of positive and negative current pulses are basically identical and repeat every one applied voltage cycle [7]. However, under certain conditions, a lot of simulation and experimental research has reported that DBDs can also work in various nonlinear modes such as asymmetric period-one (henceforth called AP1), multi-period, quasi-period and chaos through different evolution routes [8–11]. Since different discharge modes can influence the stability of DBDs in terms of the density and composition of active species in discharge plasmas and then impact their efficacy in related applications [12,13], the investigation on the control of nonlinear modes is valuable and imperative both in theoretical and practical aspects. It is worth noting that before bifurcating into other nonlinear modes from SP1, the discharge always first transforms into the asymmetric mode as a transition phase [14]. Therefore, investigating the mechanisms related to the asymmetric discharge mode could serve as a stepping-stone for the interpretation of other, more complex nonlinear behaviors in DBDs and deserve more attention.

The initiation mechanism regarding the SP1-AP1 transition was first reported by Golubovskii et al. in 2003 [15]. Through the numerical simulation of a homogeneous DBD, they observed the SP1-AP1 transition at both lower and higher concentrations of nitrogen impurities, and they concluded that such a mode transition is contributed by the residual quasi-neutral (plasma) region. Our group has also made some effort on this aspect. According to our previous works, the SP1-AP1 transition in shorter gaps might be attributed to the discordance of the evolutionary paces between electrons and ions [16], while, in longer gaps, the rise in seed electron level induced by the electron backflow is the main reason for the generation of AP1 discharge [17], and the traditional explanations such as "residual positive column" [7] or "instantaneous anode" [18] in essence are special forms of the "electron backflow region". Furthermore, we also proposed a preliminary state-controlling method which can adjust the discharge mode from AP1 to SP1 by applying a first-peak-leveled driving voltage from the beginning of the discharge [19].

Despite all those discoveries, there are still many obstacles that hinder the application of mode controlling in practical industrial applications. For example, the most favored condition in practical applications is that the discharge mode needs to switch to SP1 mode right after asymmetric discharges occur, which was not achieved in previous studies. In addition, the existing method needs to compute the initial moment of the flat-top stage, which is inconvenient for practical devices. Note that the frequency of the applied voltage can also influence the distribution of seed electrons as the amplitude of the voltage does, and changing the frequency of the power supply is much easier to implement for practical power supply devices. Therefore, in this paper, we try to use a more practical method to switch AP1 discharge into SP1 discharge by changing the driving frequency. More specifically, we first adjust the original driving frequency f_d to a different controlling frequency f_c for a certain period of time, right after the establishment of AP1 discharge. Once the discharge stabilizes under f_c, the driving frequency will be changed back to the initial value (i.e., f_d). The rest of this paper is organized as follows: Section 2 briefly introduces the one-dimensional fluid model and its qualitative validation; the numerical regulating example and its underneath mechanism are shown in Section 3; the key conclusions are drawn in Section 4; Appendix A presents the chemical scheme used in our simulation.

2. Model Description

As shown in Figure 1, the atmospheric DBD considered in the simulation is generated in a gap filled with pure helium between two parallel-plate electrodes, both covered by a thin dielectric layer with a relative permittivity of 7.5, corresponding to the value of mica glass. The upper electrode is connected to a sinusoidal voltage with an amplitude V_{am} of 2 kV, whereas the lower one is grounded. The width of the gas gap d_g is fixed to 4.4 mm and the dielectric layer thickness d_b is 1 mm. Note that the gap width is not very large (<5 mm) and much shorter compared with the radius of electrodes ($\approx$56.4 mm) [20]; in this case, the radially homogeneous assumption should be reasonable and accepted. On this basis, a one-dimensional fluid model is appropriate to simulate the discharge process, which has also been successfully applied to investigate the nonlinear phenomena in [6,21,22].

In the 1-D fluid model, electron properties, including the electron density and energy density, are governed by the Boltzmann equation under drift-diffusion approximation [23–25], as given by Equations (1)–(4); Equations (5) and (6) are multi-component transport equations [26,27] determining the flux of heavy species. Besides, in order to determine the electric field distribution, Poisson's equation is firmly coupled.

$$\frac{\partial n_e}{\partial t} + \nabla \cdot \Gamma_e = S_e \tag{1}$$

$$\frac{\partial n_\varepsilon}{\partial t} + \nabla \cdot \Gamma_\varepsilon + E\Gamma_e = S_{en} \tag{2}$$

$$\Gamma_e = -\mu_e n_e E - D_e \nabla \cdot n_e \tag{3}$$

$$\Gamma_\varepsilon = -\frac{5}{3}\mu_e n_\varepsilon E - \frac{5}{3} D_e \nabla \cdot n_\varepsilon \tag{4}$$

$$\rho \frac{\partial \omega_k}{\partial t} = \nabla \cdot J_k + S_k \tag{5}$$

$$J_k = \rho \omega_k \left(D_k \frac{\nabla \omega_k}{\omega_k} + D_k \frac{\nabla M_n}{M_n} - z_k \mu_k E \right) \tag{6}$$

Here, n_e, n_ε, Γ_e, and Γ_ε respectively denote the electron number density, electron energy, total electron flux, and electron energy flux; J_k is the flux vector for heavy species k, and its mass fraction and charge number are ω_k and z_k, respectively; S_e, S_{en}, and S_k respectively represent the source terms describing the net changing rate of electron density, electron energy loss/gain, and the source term for heavy species k; the electron mobility μ_e is solved by Bolsig$^+$ with the cross-section data from IST-Lisbon Database [28–30], then the electron diffusion coefficient D_e can be obtained through Einstein's relation [29]; the mobility and diffusion coefficient for heavy species k, i.e., μ_k and D_k, are referenced from [25]; E is the electric field intensity; the mixture properties ρ and M_n stand for the density of mixture and its mean molar mass.

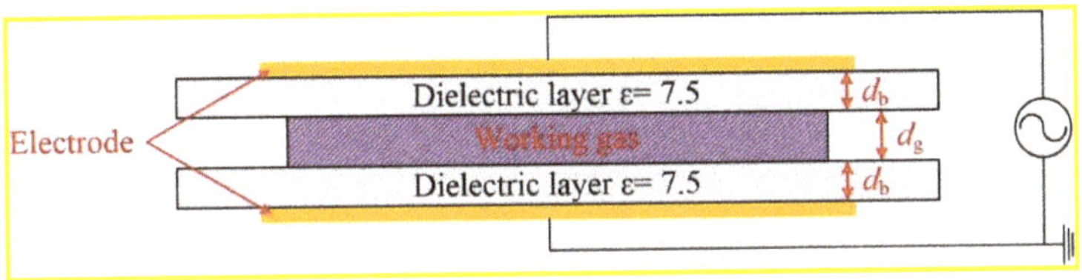

Figure 1. Schematic of the model geometry. ε denotes the relative permittivity of the dielectric layer; d_b and d_g represent the widths of the dielectric layer and gas gap, respectively.

The "wall" boundary applied at the surfaces sandwiched between plasma and dielectrics provides two sets of equations, describing the particle flux (Equations (7)–(9)) and the charge accumulation (Equations (10) and (11)), respectively.

$$\mathbf{n} \cdot \Gamma_e = \frac{1}{2} v_{e,th} n_e - \alpha_s n_e \mu_e E \cdot \mathbf{n} - \sum_i \gamma_i (\Gamma_i \cdot \mathbf{n}) \tag{7}$$

$$\mathbf{n} \cdot \Gamma_\varepsilon = \frac{5}{6} v_{e,th} n_\varepsilon - \alpha_s n_\varepsilon \mu_\varepsilon E \cdot \mathbf{n} - \sum_i \gamma_i \overline{\varepsilon_i} (\Gamma_i \cdot \mathbf{n}) \tag{8}$$

$$\mathbf{n} \cdot \Gamma_k = M_k R_{surf,k} + \alpha_s M_k c_k \mu_{k,m} z_k E \cdot \mathbf{n} \tag{9}$$

$$\frac{d\sigma_s}{dt} = J_e \cdot \mathbf{n} + J_i \cdot \mathbf{n} \tag{10}$$

$$\sigma_s = (D_2 - D_1) \cdot \mathbf{n} \tag{11}$$

Here, $\mathbf{n}$ is the unit normal vector towards the surface; both Γ_k and Γ_i stand for the boundary flux of heavy species but the latter only denotes the term for ions; $v_{e,th}$ is the thermal velocity; M_k represents the molar weight; $R_{surf,k}$ is the surface reaction rate; γ_i is the emission coefficient of the secondary electron cited from [25] and $\overline{\varepsilon_i}$ is its mean energy; σ_s represents the surface charge density on the wall, and J_e and J_i respectively denote the electron density and the ion density there; D_1 and D_2 are the electric displacement vectors on both sides of the interface; α_s is a switching function as given below.

$$\alpha_s = \begin{cases} 1, & \text{sgn}(q)E \cdot \mathbf{n} > 0 \\ 0, & \text{sgn}(q)E \cdot \mathbf{n} \leq 0 \end{cases} \tag{12}$$

Here, q represents the signed charge.

For chemical kinetics, we reused ones from our previous work [19]. To be specific, the model sustains itself under the pressure of 760 Torr and temperature of 300 K, and it considers six particles (i.e., e (electron), He, He*, He$_2$*, He$^+$ and He$_2$$^+$) and 27 reactions which are listed in Table A1 in the Appendix A. The initial densities of e, He$^+$, and He$_2$$^+$ were set to 1×10^{13} m^{-3}, 5×10^{12} m^{-3}, and 5×10^{12} m^{-3}, respectively, to ensure the initial neutral condition. The whole calculating domains were discretized through finite element method (log formulation, linear shape function), and a direct solver PARDISO of COMSOL software was employed to solve the above equations. Besides, we have taken a numerical test in terms of the initial densities and concluded that such values did not significantly influence the results in the steady state but affected the calculation time. Based on the above, a qualitative validation compared with the experimental results presented by Mangolini et al. [28] is further given, as shown in Figure 2.

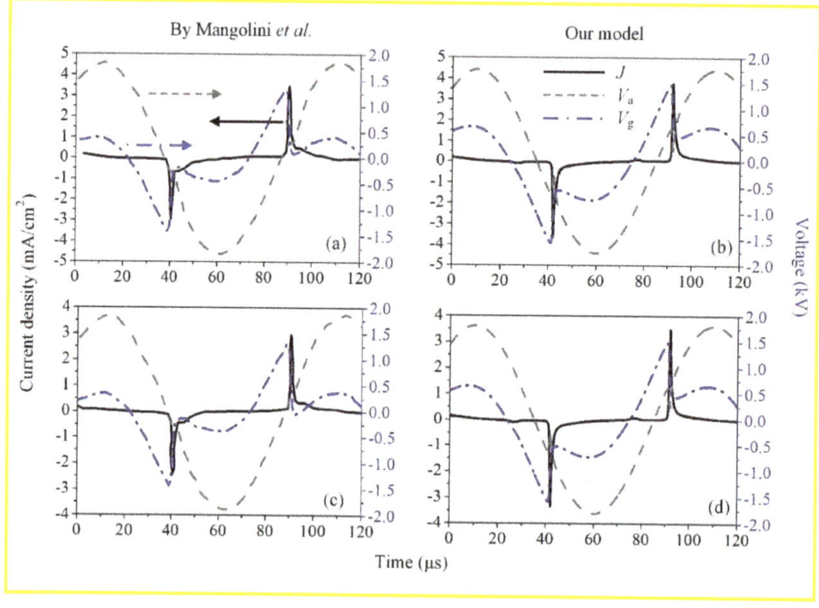

Figure 2. Comparison between (**a**) and (**c**) previous experimental waveforms and (**b**) and (**d**) their corresponding simulation results obtained from our model under mostly the same conditions. J, V_a, and V_g respectively denote the total discharge current density, applied voltage, and gap voltage. The experiment was assumed to have 100 ppm nitrogen impurities, whereas our simulation considers the pure helium. The main external parameters in (**a**) and (**b**) are: V_{am} = 1776 V, f = 10 kHz, d_g = 5 mm, d_b = 1 mm; in (**c**) and (**d**) are: V_{am} = 1813 V, f = 10 kHz, d_g = 5 mm, d_b = 1.5 mm.

Obviously, although the discharge current phase in the experiment was slightly ahead of that in our model, and the current pulse magnitude shows a difference between the experiment and simulation, the sketches of the total discharge current density (J) and gap voltage (V_g) waveforms predicted by our model principally match well with those observed in the experiment. Note that such a discrepancy in the current waveform might be contributed to the Penning ionization induced by the nitrogen impurities; that is, the extra Penning ionization causes a higher seed electron level which boosts the breakdown though this causes extinguishment more quickly. Therefore, our model is qualified to study the discharge characteristics in the homogeneous He DBD under atmospheric pressure.

3. Examples and Mechanisms of the Discharge Mode Control

3.1. SP1, AP1P, and AP1N

Before discussing the specific examples of manipulating the discharge symmetry, it is necessary to first clarify the three period-one discharge modes, i.e., SP1 discharge, asymmetric period-one discharge with a higher positive current pulse (denoted as AP1P hereinafter), and asymmetric period-one discharge with a higher negative current pulse (denoted as AP1N hereinafter), as shown in Figure 3, where the waveforms of total discharge current density under different driving frequencies are given, as well as the corresponding spatiotemporal distributions of electron density.

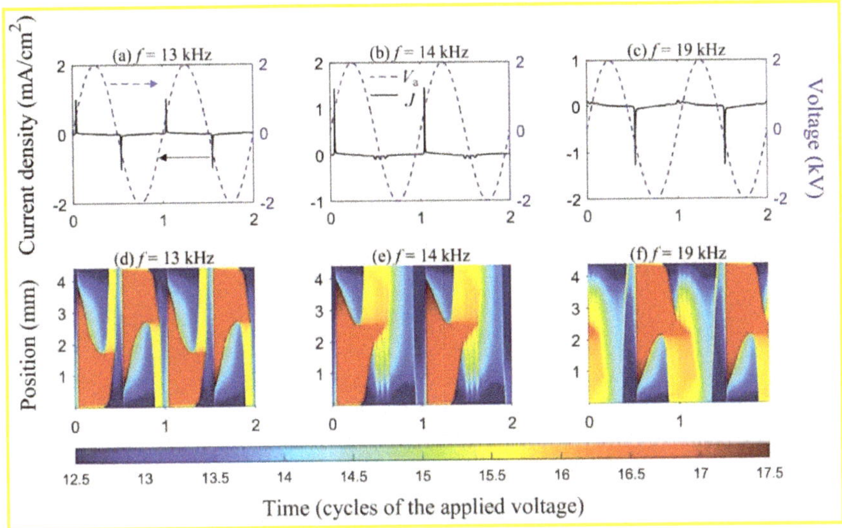

Figure 3. Waveforms of the total discharge current density (**a**–**c**) and the corresponding spatiotemporal distributions of electron density (**d**–**f**) under different driving frequencies. The voltage amplitude V_{am}, gap width d_g, and dielectric thickness d_b are fixed at 2 kV, 4.4 mm, and 1 mm, respectively.

Figure 3a shows a typical SP1 discharge where the cycles of the discharge current and the applied voltage are equal, and the shape of the discharge current in positive and negative half-cycles are also identical. Figure 3b presents an AP1P discharge whose current cycle is the same as that of the applied voltage, but the peak value of the positive current pulse is 15 times as large as that of the negative one. Contrary to AP1P, the peak value of the negative current pulse exceeds the positive current amplitude in AP1N mode, as shown in Figure 3c. Moreover, Figure 3d–f illustrate the corresponding spatiotemporal distributions of electron density in SP1, AP1P, and AP1N mode, respectively, in order to explain the reasons for the formation of three discharge modes. From Figure 3d, one can see that, before the initiation of each discharge, the residual electrons generated by the last discharge have decayed to a relatively low level. Therefore, the subsequent discharge can develop to a strong one, whose intensity is the same as that of the last discharge. Under this circumstance, the SP1 discharge mode is maintained. Figure 3e illustrates the influence of the massive residual electrons on the discharge performance in AP1P mode. With a strong positive current pulse, a large number of electrons are generated within the discharge phase. Thus, there are still a lot of electrons left in the gas gap prior to the negative discharge, which serves as the seed electrons of the subsequent negative discharge. Since a high seed electron density leads to a relatively low breakdown voltage [29], the subsequent negative discharge will start earlier, and a weak multi-pulse feature will be observed. As a result,

AP1P discharge ensues. The generation process of AP1N discharge is similar to AP1P, and the only difference lies in the polarity of the strong current pulse, as shown in Figure 3f.

3.2. Numerical Regulating Example and the Underlying Mechanism

Under the source parameters and initial values described in Section 2, the discharge under f_d of 14 kHz will finally stabilize in the AP1P mode after about 10 applied voltage cycles, as shown in Figure 4a. Starting from the 15th cycle, we change f_d to a different frequency for 20 cycles, and then adjust the frequency back to 14 kHz at the 35th cycle. The results indicate that the final stabilized discharge mode depends on f_c to some extent. Figure 4b shows that when f_c is set to 8 kHz, the discharge mode at the frequency-altered stage is SP1. However, after this stage, the discharge mode evolves back to AP1P; when f_c is set to 20 kHz, the discharge mode of the control section and the section after the 35th cycles are both AP1N, as shown in Figure 4c; when f_c increases to 30 kHz, an AP1P discharge is observed at the frequency-altered stage, but the discharge mode after the 35th cycle evolves into SP1, as shown in Figure 4d. Note that, when f_c rises beyond 30 kHz (the largest f_c considered in our simulation is 100 kHz), the initial AP1P discharge under 14 kHz can always be adjusted to SP1.

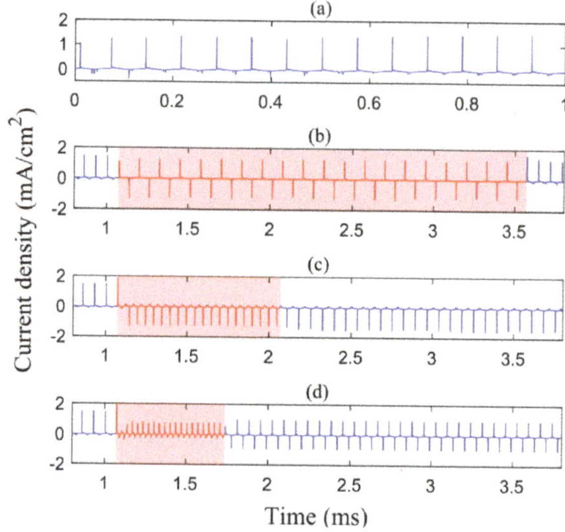

Figure 4. Temporal profiles of current density when the controlling frequency f_c is set to be (**a**) 14 kHz, (**b**) 8 kHz, (**c**) 20 kHz and (**d**) 30 kHz respectively.

At first sight, the discharge mode of the controlling section, as illustrated in Figure 4, is not directly relevant to that of the after-controlling section. To gain a deeper insight into how the frequency impacts the discharge mode, the evolution of the seed electron level has been extracted as a function of the breakdown number, as further given in Figure 5. Combined with two such figures, one can observe that, before the first breakdown of the controlling section, although the seed electron maintains at the same level (the first red circle in Figure 5), the discharge intensity and residual electron density of the subsequent breakdown show an obvious difference. This phenomenon may be qualitatively related to the equivalent gap resistance. Since the lower driving frequency causes lower angular frequency but higher equivalent gap resistance, the discharge intensity of the first breakdown in the controlling section is relatively weak when f_c is lower than f_d (14 kHz). On this basis, when f_c is set to a lower value like 8 kHz, the discharge in the controlling section develops from a relatively lower seed electron density and provides sufficient dissipative time for the residual electrons before the next breakdown, leading to an SP1 mode. The opposite leading to AP1 mode can be inferred by analogy.

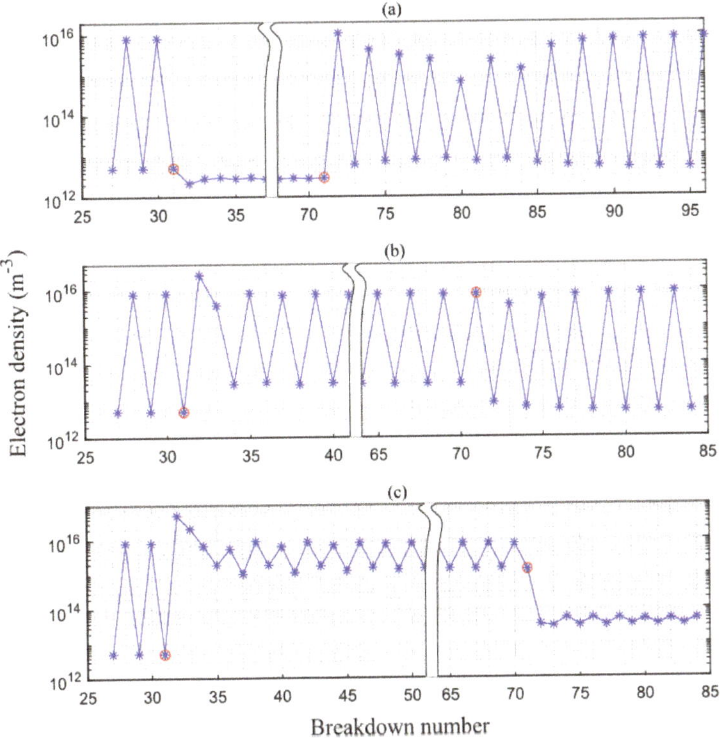

Figure 5. Evolution of the seed electron level as a function of the breakdown number, i.e., the average electron density right at the beginning moment of each breakdown. The f_c in (**a**–**c**) are 8 kHz, 20 kHz, 30 kHz, respectively. The red circles indicate the transition points of the driving frequency.

Through 20 controlling cycles, the driving frequency is adjusted back to 14 kHz. Theoretically speaking, the period of controlling section is not fixed, and the criteria considered here is to choose one relatively long period of time that ensures the discharge in the controlling section to achieve the steady state. As can be observed in Figure 5a,b, both low and high seed electron levels (the second red circle) will trigger the AP1 mode in the after-controlling section. In this case, one can understand that, in order to generate an SP1 discharge, the seed electron must retain a moderate level. Figure 6 presents the space–average seed electron relationship and the peak current density under the driving frequency of 14 kHz. In particular, the seed electron density in Figure 6a indicates the value before the breakdown, whereas the one in Figure 6b represents the residual value after the breakdown and dissipative stage. Obviously, if the seed electron density maintains at a low level at the beginning of a discharge ($<2 \times 10^{13}$ m^{-3}), then the discharge intensity of the subsequent discharge will be strong (>1.23 mA/cm^2). Correspondingly, the residual electron density after the breakdown, as further depicted in Figure 7a, can not be dissipated completely and exceeds 1×10^{17} m^{-3} (Figure 6b). Such a high initial value of electron density can only ignite an unmatured discharge, forming an AP1 mode. Analogously, if the seed electron density reaches a high level at the beginning of discharge ($>8 \times 10^{15}$ m^{-3}), then the subsequent discharge cannot develop to a mature one and will form several weak current pulses with the maximum density being less than 0.2 mA/cm^2, as shown in Figures 6a and 7b.

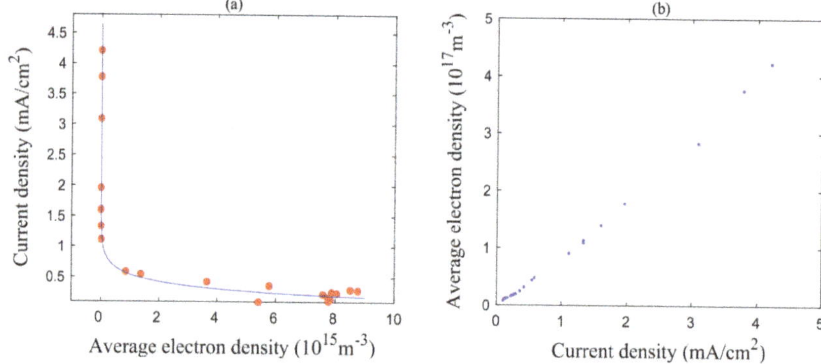

Figure 6. The relationship between the average electron density and discharge current density. (**a**) average electron density at the starting moment of the discharge vs. current density at the instant of current peak, (**b**) current density at the instant of current peak vs. maximum average electron density.

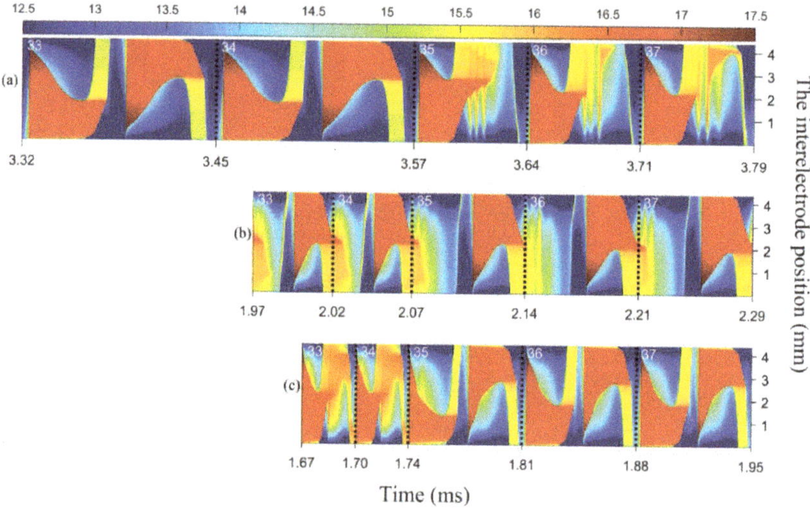

Figure 7. Spatiotemporal distribution of $\log_{10}(n_e)$ during the 33, 34, 35, 36 and 37 applied voltage cycle when f_c is set to be (**a**) 8 kHz, (**b**) 20 kHz, (**c**)30 kHz. The black dashed lines divide off different applied voltage cycles.

Based on the above analyses, the seed electron density for pursuing SP1 mode should be approximately limited in the range from 2×10^{13} to 8×10^{15} m^{-3} under our simulation conditions. Ignited by an appropriate seed electron level, the intensity of the subsequent discharge should be moderate so that the residual electrons can be dissipated completely, as shown in Figure 7c, leading to the SP1 mode. It should be pointed out that, being a control method of seed electron density, the higher driving frequency is able to limit the dissipative process of discharge. Therefore, the cooperation between seed electron level and dissipative time will lead to a less intense AP1 mode when f_c is beyond 30 kHz, as can be seen from the current waveform and seed electron evolution illustrated by Figures 4 and 5. We have carried out a parameterized sweep in terms of f_c from 30 to 100 kHz, and the results show that, within this range of frequency, such a less intense AP1 discharge can impose a restriction on the drastic variation of the seed electron density and finally, the initial density of electrons before the breakdown varies within the abovementioned range (from 2×10^{13} to 8×10^{15} m^{-3}). That is to say, increasing the driving frequency would be a practical strategy to manipulate the discharge symmetry and some more in-depth investigations would be carried out in our future study.

4. Conclusions

In this paper, we put forward a practical method for controlling the discharge symmetry of atmospheric homogeneous dielectric barrier discharges by adjusting the driving frequency. Through a qualitatively validated 1D fluid model, the discharge evolution, manipulating process, and underlying mechanism are presented. Some important conclusions are drawn as follows:

(1) The practical control strategy proposed here first changes the original driving frequency to a relatively larger one until the discharge stabilizes again, and then turns the driving frequency back to the original one;

(2) Three period-one discharge modes can be converted to each other by applying different control frequencies;

(3) The effectiveness of the control strategy is determined by the seed electron level at the frequency-altered phase, and there is a critical range of the seed electron density. Under the original driving frequency of 14 kHz, the seed electron level approximately ranges from 2×10^{13} m^{-3} to 8×10^{15} m^{-3};

(4) The higher driving frequency in the controlling section can limit the dissipative process of discharge, and further induce a less intense AP1 mode through the cooperation between seed electron level and dissipative time. In our simulations, the discharges with an initial driving frequency of 14 kHz can always be converted to SP1 mode when the control frequency is beyond 30 kHz.

Author Contributions: Conceptualization, L.L. (Ling Luo); Funding acquisition, D.D.; Methodology, Q.W. and Y.Z.; Project administration, D.D.; Supervision, L.L. (Licheng Li); Writing—original draft, L.L. (Ling Luo); Writing—review & editing, Q.W. All authors have read and agreed to the published version of the manuscript.

Funding: This work is supported by the National Natural Science Foundation of China (Grant No. 51877086).

Appendix A

Table A1. Chemical reactions considered in the model.

Index	Reaction	Rate Coefficient	Reference
R1	$e + He => e + He$	$f(T_e)$	[30]
R2	$e + He => e + He^*$	$f(T_e)$	[30]
R3	$e + He => 2e + He^+$	$f(T_e)$	[30]
R4	$e + He^* => 2e + He^+$	$1.28 \times 10^{-7} \times T_e^{0.6} \times \exp(-4.78/T_e)$	[31]
R5	$e + He^* => e + He$	2.9×10^{-9}	[31]
R6	$e + He_2^* => e + 2He$	3.8×10^{-9}	[31]
R7	$2e + He^+ => e + He^*$	$5.82 \times 10^{-20} \times (T_e/0.026)^{-4.4}$	[31]
R8	$2e + He_2^+ => He^* + He + e$	2.8×10^{-20}	[31]
R9	$e + He + He_2^+ => He^* + 2He$	3.5×10^{-27}	[31]
R10	$2e + He_2^+ => He_2^* + e$	1.2×10^{-21}	[31]
R11	$e + He + He_2^+ => He_2^* + He$	1.5×10^{-27}	[31]
R12	$e + He^+ => He^*$	$6.76 \times 10^{-13} \times T_e^{-0.5}$	[32]
R13	$e + He + He^+ => He + He^*$	$1 \times 10^{-26} \times (T_e/0.026)^{-2}$	[33]
R14	$e + He_2^+ => He^* + He$	$8.82 \times 10^{-9} \times (T_e/0.026)^{-1.5}$	[33]
R15	$e + He_2^+ => 2He$	1.0×10^{-8}	[33]
R16	$e + He + He_2^+ => 3He$	2.0×10^{-27}	[31]
R17	$e + He_2^* => 2e + He_2^+$	$9.75 \times 10^{-16} \times T_e^{0.71} \times \exp(-3.4/T_e)$	[32]
R18	$He^* + 2He => 3He$	2.0×10^{-34}	[31]

Table A1. *Cont.*

Index	Reaction	Rate Coefficient	Reference
R19	$He^* + He^* => e + He_2^+$	2.03×10^{-9}	[31]
R20	$He^* + He^* => e + He + He^+$	8.7×10^{-10}	[15]
R21	$He^+ + 2He => He_2^+ + He$	1.4×10^{-31}	[31]
R22	$He^* + 2He => He_2^* + He$	2.0×10^{-34}	[34]
R23	$He^* + He_2^* => He^+ + 2He + e$	5.0×10^{-10}	[15]
R24	$He^* + He_2^* => He_2^+ + He + e$	2.0×10^{-9}	[35]
R25	$He_2^* + He_2^* => He^+ + 3He + e$	3.0×10^{-10}	[35]
R26	$He_2^* + He_2^* =>$ $He_2^+ + 2He + e$	1.2×10^{-9}	[15]
R27	$He_2^* + He => 3He$	1.5×10^{-15}	[33]

Note: T_e denotes the electron temperature in eV. He^* represents $He(2^3 S)$ and $He(2^1 S)$. He_2^* represents $He_2(a^3\Sigma_u^+)$. n_e is the electron density in m^{-3}. The unit of reaction rate coefficient for two body reactions and three body reactions are m^3/s and m^6/s, respectively. $f(T_e)$ indicates the rate coefficient as function of the electron mean temperature calculated by Bolsig+ [30], and the cross-section data from IST-Lisbon database [36] was used as the input parameter of the calculation.

References

1. Sun, B.; Liu, D.; Iza, F.; Sui, W.; Yang, A.; Liu, Z.; Rong, M.; Wang, X. Global model of an atmospheric-pressure capacitive discharge in helium with air impurities from 100 to 10000 ppm. *Plasma Sources Sci. Technol.* **2018**, *28*, 35006. [CrossRef]

2. Shao, T.; Wang, R.; Zhang, C.; Yan, P. Atmospheric-pressure pulsed discharges and plasmas: Mechanism, characteristics and applications. *High Volt.* **2018**, *3*, 14–20. [CrossRef]

3. Zhen, Y.; Sun, H.; Wang, W.; Jia, M.; Jin, D. Thermal characterisation of dielectric barrier discharge plasma actuation driven by radio frequency voltage at low pressure. *High Volt.* **2018**, *3*, 154–160. [CrossRef]

4. Shao, T.; Yang, W.; Zhang, C.; Niu, Z.; Yan, P.; Schamiloglu, E. Enhanced surface flashover strength in vacuum of polymethylmethacrylate by surface modification using atmospheric-pressure dielectric barrier discharge. *Appl. Phys. Lett.* **2014**, *105*, 71607.

5. Wang, S.; Yang, D.; Zhou, R.; Zhou, R.; Fang, Z.; Wang, W.; Ostrikov, K. Mode transition and plasma characteristics of nanosecond pulse gas–liquid discharge: Effect of grounding configuration. *Plasma Process. Polym.* **2019**, e1900146. [CrossRef]

6. Li, X.; Liu, R.; Jia, P.; Wu, K.; Ren, C.; Yin, Z. Influence of driving frequency on discharge modes in the dielectric barrier discharge excited by a triangle voltage. *Phys. Plasmas* **2018**, *25*, 13512. [CrossRef]

7. Dai, D.; Hou, H.; Hao, Y. Influence of gap width on discharge asymmetry in atmospheric pressure glow dielectric barrier discharges. *Appl. Phys. Lett.* **2011**, *98*, 131503.

8. Zhang, D.; Wang, Y.; Wang, D. Numerical study on the discharge characteristics and nonlinear behaviors of atmospheric pressure coaxial electrode dielectric barrier discharges. *Chin. Phys. B* **2017**, *26*, 65206. [CrossRef]

9. Zhang, D.; Wang, Y.; Wang, D. The nonlinear behaviors in atmospheric dielectric barrier multi pulse discharges. *Plasma Sci. Technol.* **2016**, *18*, 826. [CrossRef]

10. Wang, Y.; Shi, H.; Sun, J.; Wang, D. Period-two discharge characteristics in argon atmospheric dielectric-barrier discharges. *Phys. Plasmas* **2009**, *16*, 63507. [CrossRef]

11. Ouyang, J.; Li, B.; He, F.; Dai, D. Nonlinear phenomena in dielectric barrier discharges: Pattern, striation and chaos. *Plasma Sci. Technol.* **2018**, *20*, 103002. [CrossRef]

12. Walsh, J.L.; Iza, F.; Janson, N.B.; Kong, M.G. Chaos in atmospheric-pressure plasma jets. *Plasma Sources Sci. Technol.* **2012**, *21*, 34008. [CrossRef]

13. Walsh, J.L.; Iza, F.; Janson, N.B.; Law, V.J.; Kong, M.G. Three distinct modes in a cold atmospheric pressure plasma jet. *J. Phys. D Appl. Phys.* **2010**, *43*, 75201. [CrossRef]

14. Zhang, D.; Wang, Y.; Wang, D. The transition mechanism from a symmetric single period discharge to a period-doubling discharge in atmospheric helium dielectric-barrier discharge. *Phys. Plasmas* **2013**, *20*, 63504. [CrossRef]

15. Golubovskii, Y.B.; Maiorov, V.A.; Behnke, J.; Behnke, J.F. Modelling of the homogeneous barrier discharge in helium at atmospheric pressure. *J. Phys. D Appl. Phys.* **2002**, *36*, 39. [CrossRef]

16. Ning, W.; Dai, D.; Zhang, Y.; Hao, Y.; Li, L. Transition from symmetric discharge to asymmetric discharge in a short gap helium dielectric barrier discharge. *Phys. Plasmas* **2017**, *24*, 73509. [CrossRef]

17. Zhang, Y.; Dai, D.; Ning, W.; Li, L. Influence of electron backflow on discharge asymmetry in atmospheric helium dielectric barrier discharges. *AIP Adv.* **2018**, *8*, 95327. [CrossRef]

18. Ha, Y.; Wang, H.; Wang, X. Modeling of asymmetric pulsed phenomena in dielectric-barrier atmospheric-pressure glow discharges. *Phys. Plasmas* **2012**, *19*, 12308. [CrossRef]

19. Zhang, Y.; Ning, W.; Dai, D. Numerical investigation on the transient evolution mechanisms of nonlinear phenomena in a helium dielectric barrier discharge at atmospheric pressure. *IEEE Trans. Plasma Sci.* **2018**, *47*, 179–192. [CrossRef]

20. Zhang, Y.; Ning, W.; Dai, D. Influence of nitrogen impurities on the performance of multiple-current-pulse behavior in a homogeneous helium dielectric-barrier discharge at atmospheric pressure. *J. Phys. D Appl. Phys.* **2018**, *52*, 45203. [CrossRef]

21. Zhang, Y.; Ning, W.; Dai, D. Numerical investigation on the dynamics and evolution mechanisms of multiple-current-pulse behavior in homogeneous helium dielectric-barrier discharges at atmospheric pressure. *AIP Adv.* **2018**, *8*, 35008. [CrossRef]

22. Zhang, Z.; Nie, Q.; Zhang, X.; Wang, Z.; Kong, F.; Jiang, B.; Lim, J. Ionization asymmetry effects on the properties modulation of atmospheric pressure dielectric barrier discharge sustained by tailored voltage waveforms. *Phys. Plasmas* **2018**, *25*, 43502. [CrossRef]

23. Yan, W.; Xia, Y.; Bi, Z.; Song, Y.; Wang, D.; Sosnin, E.A.; Skakun, V.S.; Liu, D. Numerical and experimental study on atmospheric pressure ionization waves propagating through a U-shape channel. *J. Phys. D Appl. Phys.* **2017**, *50*, 345201. [CrossRef]

24. Huang, Z.; Hao, Y.; Yang, L.; Han, Y.; Li, L. Two-dimensional simulation of spatiotemporal generation of dielectric barrier columnar discharges in atmospheric helium. *Phys. Plasmas* **2015**, *22*, 123509. [CrossRef]

25. Lazarou, C.; Belmonte, T.; Chiper, A.S.; Georghiou, G.E. Numerical modelling of the effect of dry air traces in a helium parallel plate dielectric barrier discharge. *Plasma Sources Sci. Technol.* **2016**, *25*, 55023. [CrossRef]

26. Lazarou, C.; Koukounis, D.; Chiper, A.S.; Costin, C.; Topala, I.; Georghiou, G.E. Numerical modeling of the effect of the level of nitrogen impurities in a helium parallel plate dielectric barrier discharge. *Plasma Sources Sci. Technol.* **2015**, *24*, 35012. [CrossRef]

27. Lazarou, C.; Chiper, A.S.; Anastassiou, C.; Topala, I.; Mihaila, I.; Pohoata, V.; Georghiou, G.E. Numerical simulation of the effect of water admixtures on the evolution of a helium/dry air discharge. *J. Phys. D Appl. Phys.* **2019**, *52*, 195203. [CrossRef]

28. Mangolini, L.; Anderson, C.; Heberlein, J.; Kortshagen, U. Effects of current limitation through the dielectric in atmospheric pressure glows in helium. *J. Phys. D Appl. Phys.* **2004**, *37*, 1021. [CrossRef]

29. Wang, Q.; Ning, W.; Dai, D.; Zhang, Y. How does the moderate wavy surface affect the discharge behavior in an atmospheric helium dielectric barrier discharge model? *Plasma Process. Polym.* **2019**, e1900182. [CrossRef]

30. Hagelaar, G.; Pitchford, L.C. Solving the Boltzmann equation to obtain electron transport coefficients and rate coefficients for fluid models. *Plasma Sources Sci. Technol.* **2005**, *14*, 722. [CrossRef]

31. Deloche, R.; Monchicourt, P.; Cheret, M.; Lambert, F. High-pressure helium afterglow at room temperature. *Phys. Rev. A* **1976**, *13*, 1140. [CrossRef]

32. Yuan, X.; Raja, L.L. Computational study of capacitively coupled high-pressure glow discharges in helium. *IEEE Trans. Plasma Sci.* **2003**, *31*, 495–503. [CrossRef]

33. Wang, Q.; Economou, D.J.; Donnelly, V.M. Simulation of a direct current microplasma discharge in helium at atmospheric pressure. *J. Appl. Phys.* **2006**, *100*, 23301. [CrossRef]

34. Konstantinovskii, R.S.; Shibkov, V.M.; Shibkova, L.V. Effect of a gas discharge on the ignition in the hydrogen-oxygen system. *Kinet. Catal.* **2005**, *46*, 775–788. [CrossRef]

35. Stalder, K.R.; Vidmar, R.J.; Nersisyan, G.; Graham, W.G. Modeling the chemical kinetics of high-pressure glow discharges in mixtures of helium with real air. *J. Appl. Phys.* **2006**, *99*, 93301. [CrossRef]

36. IST-Lisbon Database. Available online: https://www.lxcat.net/ (accessed on 1 January 2020).

MDPI
St. Alban-Anlage 66
4052 Basel
Switzerland
Tel. +41 61 683 77 34
Fax +41 61 302 89 18
www.mdpi.com

Applied Sciences Editorial Office
E-mail: applsci@mdpi.com
www.mdpi.com/journal/applsci